David Schellander

CFD simulations of particle laden flows

Particle transport and separation

Anchor Academic
Publishing

Schellander, David: CFD simulations of particle laden flows: Particle transport and separation. Hamburg, Anchor Academic Publishing 2013

Buch-ISBN: 978-3-95489-171-9
PDF-eBook-ISBN: 978-3-95489-671-4
Druck/Herstellung: Anchor Academic Publishing, Hamburg, 2013

Bibliografische Information der Deutschen Nationalbibliothek:
Die Deutsche Nationalbibliothek verzeichnet diese Publikation in der Deutschen Nationalbibliografie; detaillierte bibliografische Daten sind im Internet über http://dnb.d-nb.de abrufbar.

Bibliographical Information of the German National Library:
The German National Library lists this publication in the German National Bibliography. Detailed bibliographic data can be found at: http://dnb.d-nb.de

© Anchor Academic Publishing, Imprint der Diplomica Verlag GmbH
Hermannstal 119k, 22119 Hamburg
http://www.diplomica-verlag.de, Hamburg 2013
Printed in Germany

+

For my uncle
Karl-Heinz Schellander
(* 01.05.1964, + 20.12.2010)

+

Acknowledgements

Completing this book was probably one of the most challenging activities of my life. It was a great time to spend several years in the Christian-Doppler Laboratory of Particulate Flow Modelling and the Department of Particulate Flow Modelling at Johannes Kepler University of Linz, and its members will always remain dear to me.

Special thanks go to

- my wife Ulrike Schellander,

- my family,

- my uncle Karl-Heinz Schellander(+).

Furthermore, the author want to thank Evan Smuts (Department of Chemical Engineering, University of Cape Town, Africa),who thoroughly reviewed this book.

I

Preface

This book was written during my work at the CD-Laboratory of Particulate Flow Modelling at the Johannes Kepler University in Linz. Now it is the Department of Particulate Flow Modelling under the direction of Dr. Stefan Pirker, an expert for particle laden flows. If you are interested in particle laden flows and get the chance to visit the famous city Linz in Austria, then you should also take time to visit the Johannes Kepler university. The top level research team of the Department of Particulate Flow Modelling can show you some interesting experiments and results. Please contact them before: www.particulate-flow.at

If you need additional information or more details in (discrete) particle simulation techniques you should get the book of Crowe, Sommerfeld and Tsuji named **Multiphase Flows with Droplets and Particles**. It is a top level book for PhD's and researchers and a standard reference for particulate flow modelling. Maybe, here in you will miss some detailed descriptions, then I am sure you will find them in the mentioned book.

Furthermore you find a bibliography at the end where you can see the used sources including some additional standard references for particle laden flows, for discrete Lagrangian particle models and for Eulerian granular phase models.

Enjoy reading.

Abstract

The numerical hybrid model EUgran+, which is an Eulerian-Eulerian granular phase model extended with models from the Eulerian-Lagrangian model for dense rapid particulate flows, is modified to account for poly-dispersed particle diameter distributions. These modifications include the implementation of I) a new poly-dispersed drag law and of II) new particle boundary conditions distinguishing between sliding and non-sliding particle-wall collisions and III) a new implementation of the population balance equation in the agglomeration model using the Eulerian-Lagrangian approach, referred to as Bus-stop model. Further, the applicability of the EUgran+ model is extended to cover dilute to dense poly-disperse particulate flows. Furthermore, this provides an improvement in the numerical simulation of dust separation and the formation of particle strands in industrial scale cyclones. In this book, the EUgran+Poly model is validated at 3 specific cases with different mass loadings: I) poly-dispersed particle conveying in a square pipe with a $90°$ bend at low mass loading ($L = 0.00206$); II) a particle conveying case in a rectangular pipe with a double-loop at high mass loading ($L = 1.5$); III) in a vertical pipe the implementation of the agglomeration model is validated. To show the applicability of the presented models a simulation of an industrial cyclone in experimental scale is presented. The validation and application shows that considering a poly-disperse Eulerian-Eulerian granular phase improves the accordance of the simulation results with measurements significantly. Finally, the hybrid model is a good compromise for a computational efficient simulation of particulate transport and separation with different mass loading regimes.

III

Contents

1 Introduction and motivation   **1**

   1.1 Numerical simulation of particle-laden flow 4

      1.1.1 Numerical modelling of dilute flows 5

      1.1.2 Numerical modelling of dense flows 7

      1.1.3 Numerical modelling of intermediate dilute/dense particle-laden

         flows 8

   1.2 Aim of this work 9

   1.3 Organization of this book 9

2 Eulerian granular phase modelling   **11**

   2.1 Continuity equation 12

   2.2 Momentum balance 13

   2.3 Granular temperature 14

   2.4 Radial distribution function 16

   2.5 Drag coefficient and interphase momentum exchange 20

      2.5.1 Wen and Yu 20

      2.5.2 Gidaspow 21

      2.5.3 Huilin and Gidaspow 21

   2.6 Solids Stresses 21

      2.6.1 Kinetic and collisional stresses 22

      2.6.2 Frictional stresses 23

   2.7 Turbulence modelling 25

   2.8 Boundary Conditions 25

      2.8.1 Johnson and Jackson 26

      2.8.2 Li and Benyahia 27

      2.8.3 Jenkins and Louge 28

2.8.4 Schneiderbauer et. al. 30

3 Lagrangian discrete phase modelling 32

3.1 Force balance and torque balance 33

3.2 Forces on a particle 34

    3.2.1 Drag force 34

    3.2.2 Particle rotation and Magnus force 35

    3.2.3 Saffman force 36

    3.2.4 Additional forces 38

3.3 Torque 39

3.4 Turbulent fluctuations 39

3.5 Particle wall collisions 41

    3.5.1 Restitution coefficient model 41

    3.5.2 Rough wall-particle collisions 42

4 The hybrid model EUgran+Poly 45

4.1 Motivation and overview 45

4.2 Coupling and exchange forces 47

4.3 Coupling forces on the Eulerian granular phase 49

    4.3.1 Magnus force 49

    4.3.2 Particle-wall interaction 49

    4.3.3 Modified drag law for poly-dispersity 50

4.4 Coupling forces on the Lagrangian tracer particles 52

    4.4.1 Collisional particle-solid force 53

    4.4.2 Granular pressure force 53

    4.4.3 Collisional torque 54

4.5 Simulation sequence and implementation 55

5 Agglomeration 57

5.1 Simple models 59

    5.1.1 Agglomerated filling 59

    5.1.2 Linear agglomeration 59

5.2 Particle population balance equation 60

    5.2.1 Assumptions 60

    5.2.2 Collision rates 63

        5.2.2.1 Kinematic collision rate 63

        5.2.2.2 Brownian collision rate 64

        5.2.2.3 Turbulent collision rate 64

        5.2.2.4 Comparison of collision rates 68

	5.2.3	Effective collision rate	69
	5.2.4	Sticking probability	71
5.3		Bus stop model	72
	5.3.1	Implementation	73
5.4		Volume population balance model	74

6 Validation by lab-scale experiments ... **78**

6.1		Dilute poly-dispersed flow in a duct	79
	6.1.1	Boundary conditions and simulation setup	80
	6.1.2	Results and discussion	80
6.2		Mono-dispersed flow in a medium laden duct	85
	6.2.1	Boundary conditions and simulation set up	86
	6.2.2	Results and discussion	87
6.3		Agglomeration of poly-dispersed particulate flow in a vertical pipe	90

7 Application to cyclone separation ... **93**

7.1		Hybrid Model	94
	7.1.1	Boundary conditions and simulation setup	94
	7.1.2	Results and discussion	96
	7.1.3	Results and discussion for separation of limestone material	99
	7.1.4	Discussion of computational efficiency	100
7.2		Agglomeration	101

8 Conclusions and Outlook ... **103**

A Restitution coefficients are no constants ... **106**

B Computation of Lagrangian particle-wall collision ... **108**

C UDF Structure of hybrid model ... **112**

D Cyclone dimensions based on Muschelknautz theory ... **117**

E Nomenclature ... **123**

List of Figures ... **131**

List of Tables ... **132**

Bibliography ... **133**

Abbreviations

BC Boundary conditions

CFD Computational Fluid Dynamics

DEM Discrete Element Model

DPM Discrete Phase Model

DRW Discrete Random Walk Model

EUgran+ Eulerian granular hybrid model (mono-disperse)

EUgran+Poly Eulerian granular based hybrid model (poly-disperse)

NSE Navier-Stokes equations

PBE Population Balance Equation

PTE Pseudo Thermal energy

RMS Root Mean Square

RWM Random Walk Model

UDF User defined function

VOF Volume fraction

1

Introduction and motivation

Particle-laden flows are present all over the world. They can be found in our environment, in life science and in industrial processes. Sandstorms in the desert, wind blowing up leaves and dust, sediment transport in river beds are only a few examples of environmental flows. In life science blood flow can be considered as particle-laden flow, with the blood cells representing the particles. Moreover, many industrial processes involve particle-laden flows including the conveying of particles in transport pipes, the separation of dust in cyclones and filters or heating and reaction of particles in countercurrent gas flows.

Also, in cement industry, particle-laden flows are dominant. At the beginning of the cement production process, lumpy stones are milled to small particles. The resulting particle diameters are so small that the powder is prone to particle cohesion and subsequentially agglomeration. Next, this limestone powder is introduced into the top of the preheating tower (Fig. 1.1). The preheating tower resembles a system of interconnected cyclones, which should guarantee that the downward stream of particles is heated up by the countercurrent hot exhaust gas efficiently. At the same

1

Figure 1.1: Picture of a preheater tower

time any particle loss by the exhaust gas should be avoided. Finally, at bottom of the preheating tower the conditioned particles enter the calciner.

Understanding the physical behavior of this particle-laden flow is crucial for the overall process efficiency. The actual realization of the downward particle flow has a direct impact on particle losses and offgas temperature. Obviously, this depends on the specific design and geometry of the components of the preheating tower.

Commonly, three main investigation methods for the behavior of particles in a gas flow are reported in literature: analytic or empirical considerations, experiments and numerical simulations.

A lot of empirical correlations are available for conveying and separation of dust in simplified geometries. In the case of straight conveying lines the movement of solid bulk material can be estimated by simple force balances (e.g. Muschelknautz [2010]). In cyclones the separation of dust particles can be pictured by either trajectory models or by separation surface models. A review of these methods can be found in Hoffmann and Stein [2008]. Generally all these methods are restricted to simplified geometries and boundary conditions. For example, in case of cyclone

2

separation, it has been reported that the overall separation efficiency depends on the geometry and orientation of the upstream conveying line. If the upstream duct is curved shortly before the cyclone entrance the overall separation efficiency will be affected [Abrahamson et al., 2002, Pirker and Kahrimanovic, 2007]. None of the available empirical correlations are able to account for this dependency. Nevertheless, the system of interlinked cyclones in an industrial preheating tower consists of a lot of complicated geometries which can not be modelled by standard correlations. Therefore, in our case, analytical considerations can only be used for highlighting the underlying physical phenomena.

Much knowledge of the behavior of conveying and separation of powder was gained by experiments. In principle, experiments in an industrial cement production plant could provide valuable insights into the particle flow behavior of the real process. Nevertheless, in the case of the preheating tower, any experiments must be conducted in a very harsh environment. The tower is high, the gas is hot and the platforms are not weather proof. Another crucial point is that experiments should not impair the continuous cement production. On-site experiments are very time consuming, nearly always expensive and a logistical challenge. Therefore, experiments are commonly done at laboratory-scale. In this case a big challenge is that the physical flow regimes of the lab-scale experiments agree with the behavior of the real industrial process. Nevertheless, in case of the preheating tower this requirement can hardly be met. For instance, if the gas flow velocity is adjusted such that the conveying conditions are the same as in the industrial plant, the resulting centrifugal forces on the particles inside a down-scaled cyclone are orders of magnitude larger than in the corresponding industrial cyclone. Consequently, experimental findings have to be evaluated carefully prior to drawing conclusions for industrial plants. Furthermore, even lab-scale experiments are always expensive and time consuming. Therefore, experimental methods are on the one hand limited in their predictability, and on the other hand they are not flexible enough for short term design issues.

Numerical simulations are pooled under the catchphrase Computational Fluid Dynamics (CFD). They provide flexible, fast and cheap methods for predicting the behavior of particle-laden flows. The method is based on approximation for solving the well established Navier-Stokes flow equations. In principle this method can be applied to arbitrary geometries. CFD software packages are available as open source (e.g. OpenFOAM®) and as commercial products (e.g. ANSYS FLUENT). Common numerical simulation approaches are organized into Lagrangian and Eulerian methods. The Lagrangian methods compute the movement of individual particles and their interaction with the surrounding fluid. In contrast to that, Eulerian methods assume that the multitude of particles behaves as an artificial particulate phase that interacts with the fluid phase. Using well established models guarantees that simulation results are consistent. In the case of the preheating tower, dense and dilute particle-laden regions are to be expected. Especially in case of dense laden flows numerical models are still under investigations and simulation results have to be taken with care. Therefore, an in depth knowledge of the simulated process and a careful interpretation of the simulation results is important. Consequently, CFD simulations have to be checked for plausibility and they should always be validated prior to application.

1.1 Numerical simulation of particle-laden flow

In numerical simulation of particle-laden flows, the most important challenge is that different flow regimes are governed by completely different physics. Commonly particle-laden flows are classified in dilute, medium and dense flow regimes [Dartevelle, 2003]. Therefore, the decision which regime is present is judged by the volume fraction

$$\alpha_s = \frac{V_s}{V_{cell}}. \tag{1.1}$$

The volume fraction is described by the ratio of particle/solid volume V_s in a specific volume V_{cell}. The ratio has an upper threshold which is given in a packed bed of mono-dispersed spherical particles by $\alpha_{max,p} = 0.64$ [Lun et al., 1984]. Hence, there is always a minimum of 36 % fluid inside the specific volume. In numerical simulation the volume fraction can be evaluated for each computational cell. Following Elghobashi [1994] and the literature cited therein, Figure 1.2 shows the different possible particle-laden flow regimes. Dilute particulate flows are characterized by a volume fraction α_s of particles lower than 10^{-6}. In this regime only a coupling from fluid to the particles is important (one way coupling). Particle trajectories can be calculated independently. Therefore, just the kinetics of particles are taken into account. In medium-density particle flows, $10^{-6} < \alpha_s < 10^{-3}$, the influence of the particles on the surrounding fluid should be considered (two way coupling). Additionally, the increasing number of particle-particle collisions must be taken into account for $\alpha_s > 10^{-3}$ (four way coupling). With increasing volume fraction the influence of collisions on the behavior of the flow increases. In dense particle flows,

4

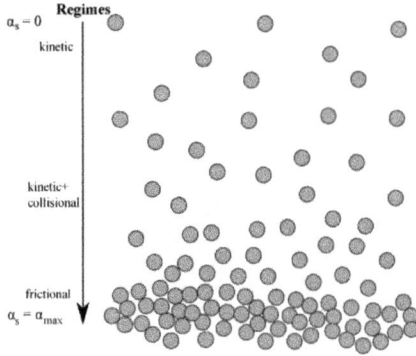

Figure 1.2: Sketch of particle-laden flow regimes

$\alpha_s > 0.5$ the friction between particles becomes important for overall flow behavior. Following this classification of particle-laden flows, methods for dilute and dense particle-laden flows have been developed. In case of the preheating tower, nearly all particle-laden flow regimes can occur within the process. Hence, a numerical simulation model which can handle dilute to dense particle flow regimes is needed.

1.1.1 Numerical modelling of dilute flows

As defined previously, in dilute gas solid flows the particle movement depends only on the flow. Hence, the influence of particles on the gas flow and inter-particle collisions are small and can be neglected. Every single particle can be considered independently by Newton's second law, given by

$$\frac{\partial m_{\mathrm{p}} \mathbf{u}_{\mathrm{p}}}{\partial t} = \sum \mathbf{F}_{p}, \tag{1.2}$$

where m_{p} denotes the mass of the particle, $\mathbf{u}_{\mathrm{p}}$ the particle velocity, t the time and $\sum \mathbf{F}_p$ the sum of forces acting on the particle. With this force balance, the motion of each particle inside the flow can be traced in time (Figure 1.3). Nevertheless, with increasing density of particles ($\alpha_s > 10^{-6}$) the method has to be extended in order to account for the influence of the particles on the gas flow. This approach is named Particle-Source-In Cell model [Crowe et al., 1977] and provides exchange terms from particles to the fluid phase. In doing so, it regards the discrete particle phase as a source of mass, momentum and energy to the fluid phase. As a next step,

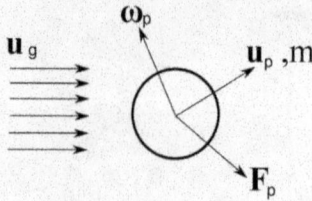

Figure 1.3: Sketch of single particle that is followed by time.

particle-particle interactions should be included to the simulation model. This is especially important if the particles under consideration are prone to agglomeration. There are two possibilities to achieve this. The collisions can be included either by a stochastic consideration [Oesterlé, 1993] or by deterministic methods [Cundall and Strack, 1979]. Stochastic methods compute the collisions of a particle with virtual collision partners based on the probability for a collision. The stochastic method was developed by Oesterlé [1993] and extended and validated by Sommerfeld [1996], Frank [2002] and Hussmann et al. [2007]. A brief introduction and comparison of the models can be found in Rao and Nott [2008]. Deterministic models resolve each collision of a particle by searching for real collision partners in the near region of the particle. Simulation methods using the deterministic collision detection are named Discrete Element Methods (DEM). For example, this method is used in the open source code LIGGGHTS [Kloss, 2011]. This DEM code has been coupled to OpenFOAM to account for the interstitial gas flows [Goniva et al., 2012]. This coupling is distributed at www.cfdem.com. An advantage of the discrete phase model is that every impact force on a particle is resolved. Additionally, it is straight forward to include particle rotation (Magnus force) and to model particle reflection at walls. However, in the case of DEM the computational efforts rise by the square of the number of particles. Hence, the approach is limited by computational resources, even new computers can only handle up to few million particles. Finally, with respect to the preheating tower, the DEM method will fail due to an excess number of particles inside the computational domain. Even in case of the stochastic approach there are still to many particles involved. Therefore, this case of dense particle-laden flow can not be simulated with a pure Lagrangian modelling approach.

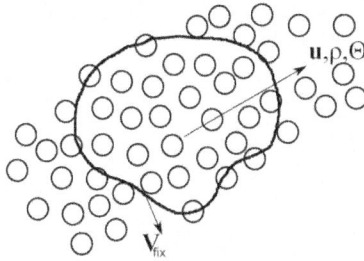

Figure 1.4: Sketch of granular flow through fixed control volume

1.1.2 Numerical modelling of dense flows

As mentioned before, dense particle-laden flows depend on the gas flow, particle-particle collisions and for flows near the packing bed volume fraction, the friction between particles. Obviously, this kind of flow contain an immense number of particles. In many cases it seems as they act like an independent phase. Hence, a concept named particular phase was developed and introduced [Ishii, 1975]. This method considers the particles as a continuous medium. Hence, the discrete particle properties are replaced by quantities of velocity, density and volume fraction for the particle phase. These quantities are assumed to be smooth functions of position and time. Therefore, a mass and force balance can be built for the particle medium inside a fixed control volume [Agrawal et al., 2001] (Figure 1.4). With this approach a multi-phase Navier-Stokes equation model can be taken to model granular flows [Gidaspow, 1994]. Thus, specific correlations for granular temperature and granular stresses have been deduced from kinetic theory. Furthermore, a momentum exchange term between the fluid and granular phases has to be included. An introduction to this model, including a discussion of its sub-models, can be found in Gidaspow [1994], Brilliantov, N. V. and Pöschel, T. [2004] and Rao and Nott [2008]. The advantage of this model is that for each phase just a set of balance equations has to be solved. In case of a two-phase gas and solid flow the computational effort is rather low. This continuum model is a good choice for particle-laden flows which are dominated by inter-particle collisions. Negative aspects of this model are that the Magnus force and realistic particle-wall interactions are currently not included. Furthermore, poly-dispersed flows can only be handled by computing a continuum particulate phase for each diameter. This would provide the chance to include the famous population balance equation of Smoluchowski [1917] for the computation of agglomeration, but as a consequence, a new set of balance equations must be solved

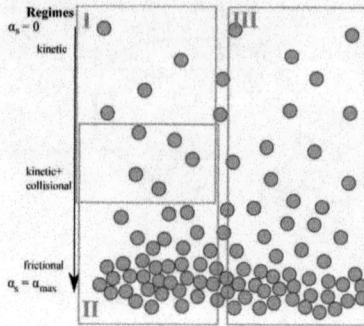

Figure 1.5: Sketch of particle-laden flow regimes; depicting the applicability of (I) Lagrangian particle tracing, (II) Eulerian particulate phase model and (III) hybrid particle-laden flow modelling

for each particle diameter. This increases the computational effort dramatically. In the case of the preheating tower example, the continuous phase method seems to be a good choice. The model counts for many of the physical effects in the conveying and separation of dense particle-laden flows. But, there are still some questions remaining regarding the Magnus force, the particle-wall treatment and poly-dispersed particle-laden flow handling.

1.1.3 Numerical modelling of intermediate dilute/dense particle-laden flows

The transport of powder, as it is observed in the preheating tower, contains a huge number of small particles which are arranged in dilute and dense particle-laden flow situations. While in some regions strand formations are observed, in other regions a dispersed flow is present. This calls for an approach that considers regime changes in particle-laden flows. To achieve this, a combination of dilute and dense flow models can be used (Figure 1.5). Two methods are conceivable, an embedded method [Zwinger, 2000, Pirker and Kahrimanovic, 2007] and a joint domain method [Pirker et al., 2010]. In the embedded method the computational domain is divided into two distinct regions. In dilute laden regions it computes the particulate flow with a numerical model for dilute flows. In all other regions a numerical model for dense flow is used. For this approach the volume fraction α_s in the geometry must be estimated a priori. In contrast to this, the joint domain model computes both dilute and dense particle flow approaches in the whole domain and weights the impact of each method according to the actual volume fraction. Remembering the preheating tower, the positions of dilute or dense flow regimes are not known a priori. Therefore, a joint method should be taken for the simulation of this industrial application. For reasons of simplicity, this joint domain combination will be abbreviated as **hybrid model** throughout this book. These hybrid models are currently under development and intended for modelling industrial applications containing particle-laden flows, where different flow regimes are present.

1.2 Aim of this work

This study aims at further developing a joint domain hybrid model for picturing particle-laden flows with locally and time dependent changing flow regimes. Therefore, special emphasis should be given to poly-dispersity, particle-wall boundary conditions and particle rotation. Furthermore, a concept for incorporating the effect of agglomeration should be suggested as an extension of the developed hybrid model. Finally, this hybrid model will be applied to particle-laden flows in the cement industry. In industrially used numerical simulation codes the numerical stability is important. A modular concept should be used in the design and programming of the method. Then, there should be the possibility to change individual sub-models in a simple way. Finally, the model should be open for further extensions in future, e. g. including the consideration of heat transfer and chemical reactions.

1.3 Organization of this book

This book is organized in three main steps: model development, model validation and model application (Figure 1.6). To guarantee a good readability this book starts with basic information on model development, which consists of four chapters. In Chapter 2 the numerical modelling of dense particle-laden flows with the Eulerian granular phase model is presented. Later on in Chapter 3 a numerical model for single particle trajectory calculation - the Euler-Lagrangian discrete phase model - is shown. In Chapter 4 a hybrid model for poly-dispersed granular flows, entitled EUgran+Poly, is presented. Furthermore, in Chapter 5, a concept study on agglomeration modelling is presented including a new implementation method of the population balance equation by Smoluchowski [1917]. Later on in Chapter 6 the hybrid and agglomeration models are tested by three validation examples. First, based on the research study of Mohanarangam et al. [2007], the hybrid model is validated for the poly-dispersed, dilute particle flow case in a duct bend. Secondly, based on the work of Pirker et al. [2010], the hybrid model is validated for the mono-dispersed, dense particle-laden flow case in a curved rectangular duct. The plausibility of the agglomeration model is demonstrated by a poly-dispersed particle-laden downpipe. In this example the agglomeration of particles during sedimentation is simulated. The presented implementation of the agglomeration model is compared to the standard implementation of the population balance equation by Smoluchowski [1917]. Finally, in Chapter 7 the hybrid model is applied to a real cyclone geometry. The cyclone, designed with theory of Muschelknautz et al. [1994] and Hoffmann and Stein [2008] was created at the Johannes Kepler university, CD-Laboratory on particulate flow's experimental facility [Krainz, 2007]. Simulations with the EUgran+Poly model are compared to measurements and analytical results of the cyclone. Additionally, an industrial cyclone at experimental scale is simulated and compared to measured results. In this application the impact of the new developed sub-models in the hybrid model can be highlighted. The book is closed by conclusions and an outlook. In the appendix, additional information such as: restitution coefficients, the modelling of the cyclone with Muschelknautz's theory, notations and curriculum vitae are given.

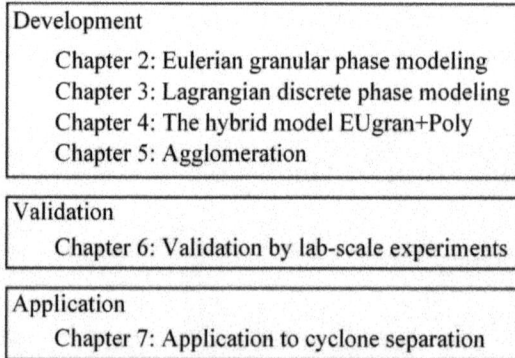

Development

    Chapter 2: Eulerian granular phase modeling
    Chapter 3: Lagrangian discrete phase modeling
    Chapter 4: The hybrid model EUgran+Poly
    Chapter 5: Agglomeration

Validation

    Chapter 6: Validation by lab-scale experiments

Application

    Chapter 7: Application to cyclone separation

Figure 1.6: Organization of this book

2

Eulerian granular phase modelling

As mentioned in the introduction, the total number of particles involved in most practically relevant particulate flows is extremely large. Hence, it may be impractical to solve the equations of motion for each particle [Agrawal et al., 2001, Schneiderbauer et al., 2012a]. It is therefore common to average the equations of motion of the individual particles, as for example in the Eulerian-Eulerian granular phase model. The model describes the flow of solid particles as a fluidized medium [Ishii, 1975, Agrawal et al., 2001]. The quantities describing the flow, e.g. velocity, density and volume fraction, are assumed to be smooth functions of position and time [Rao and Nott, 2008]. The field of application of the Eulerian granular phase model is shown in Figure 2.1. With this approach multi-phase Navier Stokes equations (NSE) are modified to account for the granular flow [Gidaspow, 1994]. This approach requires the definition of granular equivalents of temperature, pressure and stresses. For granular phases, a granular temperature (Section 2.3), a solids pressure

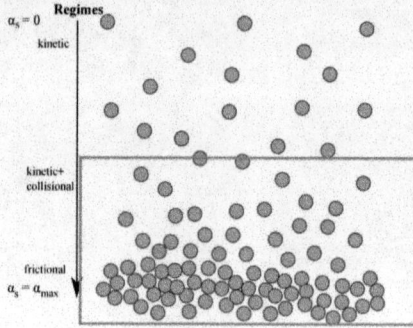

Figure 2.1: Range of application of Eulerian-Eulerian granular phase model.

and a solid shear stress tensor (Section 2.6) are included. For example, the solids shear stresses consist of a kinetic and a collisional contribution and a frictional part. Each of them contains a viscosity, which is the kinetic viscosity divided into a shear and bulk term. An overview of the Eulerian-Eulerian granular phase model and the models used therein is given in Figure 2.2.

2.1 Continuity equation

In a closed system, energy and mass is conserved. Hence, in standard fluid flows [Crowe et al., 1998] - without considering chemical reactions - a mass balance, referred to continuity equation, can be defined. In case of compressible flows for phase q it is written as

$$\frac{\partial}{\partial t}\alpha_q \rho_q + \nabla \cdot (\alpha_q \rho_q \mathbf{u}_q) = 0, \qquad (2.1)$$

where α_q denotes the volume fraction, ρ_q the density and $\mathbf{u}_q$ the velocity. The continuity equation shows that mass entering a region increases the density. In case of incompressible flows, e. g. $\rho = $ const., the equation can be simplified to

$$\frac{\partial}{\partial t}\alpha_q + \nabla \cdot (\alpha_q \mathbf{u}_q) = 0. \qquad (2.2)$$

12

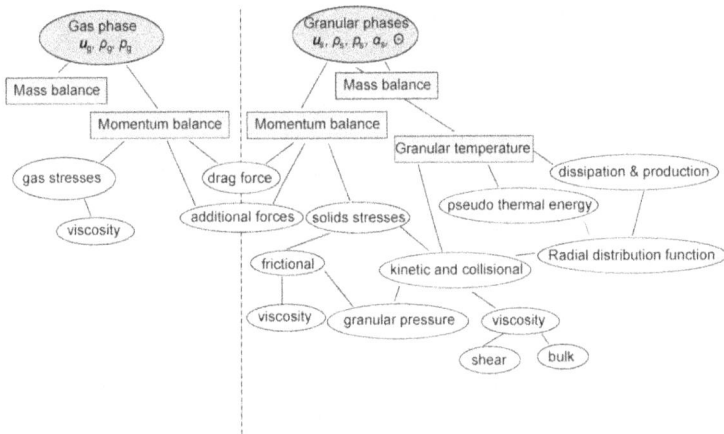

Figure 2.2: Schematic of Eulerian granular phase model, without considering any boundary conditions. The grayed ellipse describes the gas or granular phase. Inside each ellipse the computed variables are shown. Each phase model is described by a set of balancing equations which are represented by the grayed rectangles. Further models, represented by white ellipses, provide information for the balancing equations.

2.2 Momentum balance

The fluid momentum balance for multi-phase flows is nearly the same as in case of a single phase flow. It is extended by a force exchange term $\beta \left(\mathbf{u}_\mathrm{g} - \mathbf{u}_\mathrm{s} \right)$ between the particle and the gas phase and the volume fraction α_g of the phase. The momentum balance is given by

$$
\frac{\partial}{\partial t} \left(\alpha_\mathrm{g} \rho_\mathrm{g} \mathbf{u}_\mathrm{g} \right) + \nabla \cdot \left(\alpha_\mathrm{g} \rho_\mathrm{g} \mathbf{u}_\mathrm{g} \mathbf{u}_\mathrm{g} \right) =
$$
$$
- \alpha_\mathrm{g} \nabla p - \nabla \cdot \alpha_\mathrm{g} \mathbf{T}_\mathrm{g} + \beta \left(\mathbf{u}_\mathrm{g} - \mathbf{u}_\mathrm{s} \right) + \alpha_\mathrm{g} \rho_\mathrm{g} \mathbf{g} + \mathbf{f}_\mathrm{g,add}. \quad (2.3)
$$

where $\mathbf{T}_\mathrm{g}$ denotes the shear stress tensor for the gas phase (see also 2.6) and is given by,

$$
\mathbf{T}_\mathrm{g} = 2 \mu_\mathrm{g} \mathbf{D}_\mathrm{g}. \quad (2.4)
$$

13

where μ_g denotes the gas viscosity and $\mathbf{D_g}$ the rate of strain tensor, which is given for phase q as

$$\mathbf{D_q} = \frac{1}{2}\left(\nabla\mathbf{u_q} + (\nabla\mathbf{u_q})^{\mathrm{T}}\right). \qquad (2.5)$$

Furthermore, the momentum balance for a solid phase s is given by [Ding and Gidaspow, 1990, Schneiderbauer et al., 2012a],

$$\frac{\partial}{\partial t}\left(\alpha_s\rho_s\mathbf{u_s}\right) + \nabla\cdot\left(\alpha_s\rho_s\mathbf{u_s}\mathbf{u_s}\right) =$$
$$- \alpha_s\nabla p - \nabla\cdot\left(\mathbf{S}_s^{\mathrm{kc}} + \mathbf{S}_s^{\mathrm{fr}}\right) + \beta\left(\mathbf{u_g} - \mathbf{u_s}\right) + \alpha_s\rho_s\mathbf{g} + \mathbf{f}_{s,\mathrm{add}}, \qquad (2.6)$$

where $\mathbf{S}_s^{\mathrm{kc}}$ denotes the solids stress tensor arising from the kinetic and collisional contributions, $\mathbf{S}_s^{\mathrm{fr}}$ the stress tensor from frictional contributions and $\mathbf{f}_{s,\mathrm{add}}$ additional forces. With respect to the formulation of the hybrid particle model, only one particulate phase in the Eulerian granular phase model is used in this study.

2.3 Granular temperature

Corresponding to the thermal fluctuations of gas molecules a granular temperature Θ, which corresponds to the fluctuations of the particle velocities, is introduced for the granular phase. This granular temperature is defined by

$$\Theta = \frac{\langle(u - \langle u\rangle)^2\rangle}{D}, \qquad (2.7)$$

where D denotes the number of dimensions, u the velocity and $\langle u\rangle$ the average granular velocity [Goldhirsch, 2008]. The balance equation for the granular temperature is given by

$$\frac{3}{2}\left(\frac{\partial}{\partial t}\left(\alpha_s\rho_s\Theta\right) + \nabla\cdot\left(\alpha_s\rho_s\mathbf{u_s}\Theta\right)\right) = -\mathbf{S}_s^{\mathrm{kc}} : \nabla\mathbf{u_s} - \nabla\cdot\mathbf{q} - D_\Theta - \gamma_\Theta + \phi_q. \qquad (2.8)$$

14

The first term $-\mathbf{S}_{\mathrm{s}}^{\mathrm{kc}} : \nabla \mathbf{u}_{\mathrm{s}}$ denotes the generation of pseudo-thermal energy (PTE). The second term $\nabla \cdot \mathbf{q}$ represents the diffusion of pseudo-thermal energy $\mathbf{q}$, which is given by [Schneiderbauer et al., 2012a]

$$\mathbf{q} = -\frac{\kappa^*}{g_0} \left\{ \left(\frac{1}{1 + \frac{l_s}{L_C}} + \frac{12}{5} \eta_s \alpha_s g_0 \right) \left(1 + \frac{12}{5} \eta_s^2 \left(4\eta_s - 3 \right) \alpha_s g_0 \right) \right.$$
$$\left. + \frac{64}{25\pi} \left(41 - 33\eta_s \right) \eta_s^2 \alpha_s^2 g_0^2 \right\} \nabla \Theta. \quad (2.9)$$

with

$$\kappa^* = \frac{\kappa}{1 + \frac{6\beta\kappa}{5(\alpha_s\rho_s)^2 g_0\Theta}}, \quad (2.10)$$

where κ is given by

$$\kappa = \frac{75\rho_s d_s \sqrt{\pi\Theta}}{48\eta_s \left(41 - 33\eta_s \right)} \quad (2.11)$$

and

$$\eta_s = \frac{1}{2} \left(1 + e_{\mathrm{p}} \right) \quad (2.12)$$

where e_{p} denotes the particle-particle restitution coefficient. g_0 denotes the radial distribution function to include the maximum packing limit (Chapter 2.4). L_C denotes the characteristic length scale of the actual physical system [Hrenya and Sinclair, 1997] and l_s is the particle mean free path given by

$$l_s = \frac{d_s}{6\sqrt{2}g_0\alpha_s}. \quad (2.13)$$

Hence, the term $\frac{l_s}{L_C}$ recognizes the presence of boundaries [Schneiderbauer et al., 2012a]. D_Θ is the dissipation of PTE due to particle-particle collisions and is given by [Lun et al., 1984]

$$D_\Theta = \frac{48}{\sqrt{\pi}} \eta_s \left(1 - \eta_s \right) \frac{\rho_s \alpha_s^2}{d_s} g_0 \Theta^{3/2}. \quad (2.14)$$

γ_Θ describes the dissipation of PTE by viscous damping [Gidaspow et al., 1992, Gidaspow, 1994], given by

$$\gamma_\Theta = 3\beta_s \Theta. \quad (2.15)$$

15

where β_s denotes the interphase momentum exchange (e. g. Chapter 2.5). ϕ_s describes the PTE production by gas-particle slip between the solid phase and the gas phase [Koch, 1990]

$$\phi_s = \frac{81\alpha_s\mu_g^2\|\mathbf{u}_g - \mathbf{u}_s\|}{g_0 d_s^3 \rho_s \sqrt{\pi\Theta}}. \tag{2.16}$$

In dense systems the balance equation for granular temperature can be simplified to an equilibrium algebraic equation by neglecting convection and diffusion terms [van Wachem et al., 2001]. This simplified algebraic equation is used as the standard method in the commercial code FLUENT/ANSYS. Furthermore, a maximum threshold should be included to avoid discontinuities and instabilities in the granular temperature field.

2.4 Radial distribution function

The radial distribution function g_0 is a correction factor to incorporate the maximum packing limit $\alpha_{s,max}$ into the collision integral in the Boltzmann equation (e. g. Brilliantov, N. V. and Pöschel, T. [2004] and Schneiderbauer et al. [2012a]). At positions with low volume fraction $\alpha_s = 0$, the radial distribution function tends to $g_0 = 1$ and has no impact on the collision integral. In the dense regime, where α_s approaches $\alpha_{s,max}$, the radial distribution function tends to infinity ($g_0 \to \infty$), because the number of collisions tends to infinity. The maximum volume fraction $\alpha_{s,max}$ for an assembly of identical spheres is about 0.74 (e. g. named the Kepler problem) and defines the theoretical limit for a packing of mono-dispersed spheres. A detailed discussion of the hexagonal packing of spheres can be found in Conway and Sloane [1993]. In real world applications the theoretical value will never be reached and the maximum packing limit is about $\alpha_s = 0.64$. This is named the random closest packing for a mono-dispersed particle packing. The radial distribution function g_0 is described with the average particle surface to surface distance s and the distance of particle center of gravity. It is given by

$$g_0(s) = \frac{l}{s} = \frac{s + d_s}{s}. \tag{2.17}$$

To derive an expression for g_0 a model for the packing of the particles has to be assumed (e. g. Figure 2.3). In Figure 2.3 the particle packing assumption for the model derived by Sinclair and Jackson [1989] and Gidaspow and Huilin [1998] is

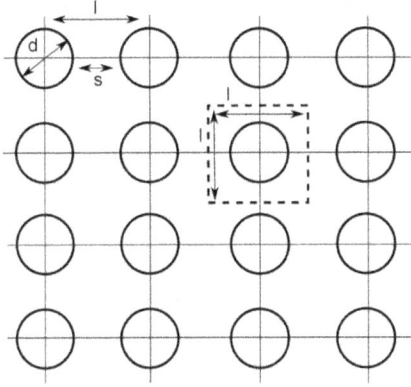

Figure 2.3: Simple model of particle packing [Sinclair and Jackson, 1989, Gidaspow and Huilin, 1998]

shown. It shows that the particles are arranged on a cubic grid. With this assumption g_0 can be described with α_s and $\alpha_{s,max}$. The ratio of d_s and l in volumetric consideration can be used to calculate the volume fraction given by

$$\alpha_s = \frac{d_s^3 \pi}{6} \frac{1}{l^3} = \frac{\alpha_{s,max} d_s^3}{l^3}, \qquad (2.18)$$

where $\alpha_{s,max} = \pi/6$, which leads to

$$\left(\frac{\alpha_s}{\alpha_{s,max}}\right)^{\frac{1}{3}} = \frac{d_s}{l} \qquad (2.19)$$

and l is calculated to

$$l = \left(\frac{\alpha_{s,max}}{\alpha_s}\right)^{-\frac{1}{3}} d_s. \qquad (2.20)$$

Recognizing that $s = l - d_s$ yields

$$s = \left(\left(\frac{\alpha_{s,max}}{\alpha_s}\right)^{-\frac{1}{3}} - 1\right) d_s. \qquad (2.21)$$

Inserting this into equation (2.17) gives [Sinclair and Jackson, 1989]

$$g_0 = \frac{s + d_s}{s} = \frac{1}{1 - \left(\frac{\alpha_{s,max}}{\alpha_s}\right)^{\frac{1}{3}}}. \qquad (2.22)$$

17

Table 2.1: Radial distribution functions

Model	Equation
Carnahan and Starling [1969]	$g_0 = \frac{1}{1-\alpha_s} + \frac{3\alpha_s}{2(1-\alpha_s)^2} + \frac{\alpha_s^2}{2(1-\alpha_s)^3}$
Lun and Savage [1986]	$g_0 = \left[1 - \frac{\alpha_s}{\alpha_{s,max}}\right]^{-2.5\alpha_s^{max}}$
Sinclair and Jackson [1989]	$g_0 = \left[1 - \left(\frac{\alpha_s}{\alpha_{s,max}}\right)^{1/3}\right]^{-1}$
Gidaspow and Huilin [1998]	$g_0 = \frac{3}{5}\left[1 - \left(\frac{\alpha_s}{\alpha_{s,max}}\right)^{1/3}\right]^{-1}$
Iddir and Arastoopour [2005]	$g_0 = \left[1 - \frac{\alpha_s}{\alpha_{s,max}}\right]^{-1}$

However, the assumption that particles will arrange on a cubic grid may be inappropriate. The measured data of Gidaspow and Huilin [1998] show that equation (2.22) underestimates g_0 in regions with low α_s. It is clear that with other assumptions on the particle packing or direct measurements of g_0 different mathematical descriptions for g_0 are found. Common models for the radial distribution function are listed in Table 2.1 and compared to measurements and computational simulations [van Wachem et al., 2001] in Figure 2.4. At the limits $\alpha_s = 0$ and $\alpha_s = \alpha_{s,max}$ most of the models approach the correct value for g_0. At $\alpha_s = 0$ the model of Gidaspow approaches $g_0 = 3/5$, whereas the correct value should be one. The model of Carnahan and Starlin gives for $\alpha_s \to \alpha_{s,max}$, $g_0 \ll \infty$. In the intermediate region of α_s, all radial distribution functions differ from measurements and an enhancement seems to be possible [Schellander et al., 2011]. This difference can be explained by the behavior of the particles, which do not tend to create a grid array, but rather form heterogeneous structures. Recognizing this, higher values for g_0 at low α_s as given in the measurements are expected. Later in this book this phenomena will be considered again in case of particle strand formation in cyclones. However, we propose to use a piecewise defined function for g_0. This function is fitted to the experiment data from Gidaspow and Huilin [1998], the computational data from Alder

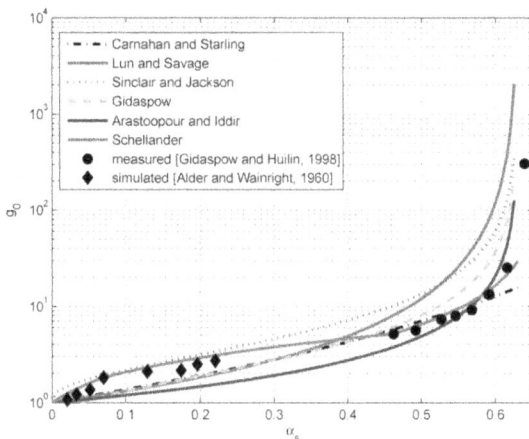

Figure 2.4: Comparison of the different models for the radial distribution function and measurements

and Wainwright [1960] and model for the radial distribution function presented by Gidaspow and Huilin [1998]. The function is defined as

$$
g_0 = \begin{cases} 1 + \frac{23}{2}\alpha_s & \alpha_s < 0.07 \\ \frac{121}{100} + \frac{17}{2}\alpha_s & 0.07 \leq \alpha_s < 0.46 \\ \frac{3}{5}\left[1 - \left(\frac{\alpha_s}{\alpha_{s,\max}}\right)^{(1/3)}\right]^{-1} & 0.46 \leq \alpha_s < 0.67 \end{cases}. \tag{2.23}
$$

In this equation the derivative of the resulting function is not completely smooth. Nevertheless, test-simulations proved that this does not impair the simulation stability. Additionally, it is assumed that the maximum packing limit is $\alpha_{s,\max} = 0.67$. Considering poly-dispersity of the granular material, which increases the maximum packing limit this assumption can be justified. If a model based on common radial distribution models should be used, a mix between Carnahan and Starling [1969] and Iddir and Arastoopour [2005] is suggested [Schneiderbauer et al., 2012a, Schneiderbauer and Pirker, 2012b],

$$
g_0 = \min\left[\left[1 - \frac{\alpha_s}{\alpha_s^{\max}}\right]^{-1}, \frac{1}{1 - \alpha_s} + \frac{3\alpha_s}{2\left(1 - \alpha_s\right)^2} + \frac{\alpha_s^2}{2\left(1 - \alpha_s\right)^3}\right]. \tag{2.24}
$$

For the simulation of fluidized beds and particle bin discharge, equation (2.24) proves to apply well. For the case of the preheating tower, where particle strand building is expected, the new piecewise defined function is suggested.

2.5 Drag coefficient and interphase momentum exchange

Physically, the fluid phase exerts several forces on the particle, e. g. drag force, Magnus force, Saffman force. The most dominant force is the drag on the particle. Hence, only the drag is considered for a coupling between the phases. Many drag laws have been proposed in the literature. In general, the interphase momentum exchange by drag is written as

$$\mathbf{f}_{s,D} = \beta_s \left(\mathbf{u}_g - \mathbf{u}_s \right) \tag{2.25}$$

where β_s is described by several models. Note that, the following models for the interphase momentum exchange are based on the assumption of homogenously distributed particles with the same diameter d_s. Three commonly used drag laws for the simulation of particle-laden flows are the models of Wen and Yu [1966], Gidaspow et al. [1992] and Huilin et al. [2003]. In dense regime, e. g. for a flow through a fixed packed bed of particles, the model of Ergun [1952] is commonly used.

2.5.1 Wen and Yu

The model of Wen and Yu [1966] is valid for particle-laden flows up to volume fractions of $\alpha_s = 0.6$, where β_s is given by

$$\beta_s = \frac{3}{4} C_D \frac{\alpha_s \alpha_g \rho_g \left| \mathbf{u}_g - \mathbf{u}_s \right|}{d_s} \alpha_g^{-2.65}, \tag{2.26}$$

with the drag coefficient C_D

$$C_D = \begin{cases} \dfrac{24 \left(1 + 0.15 (\alpha_g \mathrm{Re}_p)^{0.687} \right)}{\alpha_g \mathrm{Re}_p} & \mathrm{Re}_p \leqslant 1000 \\ \\ 0.44 & \mathrm{Re}_p > 1000 \end{cases}. \tag{2.27}$$

2.5.2 Gidaspow

For dense particulate flows Gidaspow et al. [1992] suggested a combination of the Wen and Yu correlation with the Ergun equation [Ergun, 1952]. For $\alpha_s < 0.2$ β is set to equation (2.26) and for $\alpha_s \geq 0.2$ the Ergun equation is used

$$\beta_s = 150 \frac{\alpha_s^2 \mu_g}{\alpha_1 d_s^2} + 1.75 \frac{\rho_g \alpha_s |\mathbf{u}_s - \mathbf{u}_g|}{d_s}. \tag{2.28}$$

One problem of this model is the discontinuity at the switching position of the two combined models. For the case of the preheating tower, where the complete range of α_s is present, a model like this should be used to get correct results for packed bed regions.

2.5.3 Huilin and Gidaspow

To get a smooth switching between the drag given by the Ergun equation and the Wen and Yu model, Huilin et al. [2003] have rewritten the model of Gidaspow as follows

$$\beta_s = \psi \beta_{s,\text{Ergun}} + (1 - \psi) \beta_{s,\text{WenYu}} \tag{2.29}$$

where ψ is defined as

$$\psi = \frac{1}{2} + \frac{\arctan (262.5 (\alpha_s - 0.2))}{\pi}. \tag{2.30}$$

2.6 Solids Stresses

The stress tensor for the granular phase consists of two parts. On one hand it contains the stresses arising from kinetic and collisional contributions and on the other hand the frictional stresses. These stresses can be calculated independently and the resulting solid stress is assumed to be the summation of the kinetic, collisional and frictional stresses. The closure of the system with modelling solid stresses is done with the model of Hrenya and Sinclair [1997]. It is analogous to kinetic theory of gases, based on a pseudo-thermal energy balance of the velocity fluctuations [Schneiderbauer and Pirker, 2012a].

2.6.1 Kinetic and collisional stresses

The stress tensor, arising from kinetic and collisional contributions for the solid-phase is given by

$$\mathbf{S}_s^{kc} = \left(p_s^{kc} - \lambda_s^{kc} \text{tr} \left(\mathbf{D}_s \right) \right) \mathbf{I} - 2\mu_s^{kc} \text{dev} \mathbf{D}_s, \tag{2.31}$$

where p_s^{kc} denotes the solid pressure and λ_s^{kc} the bulk viscosity. $\text{dev}\mathbf{D}_s$ denotes the deviatoric part of $\mathbf{D}_s$, which is defined as

$$\text{dev}\mathbf{D}_s = \mathbf{D}_s - \frac{1}{3}\text{tr}\left(\mathbf{D}_s\right)\mathbf{I}. \tag{2.32}$$

The shear viscosity arises from translational kinetic motion and the collisional particle-particle interactions. The solids viscosity μ_s^{kc} in a granular medium is in general decomposed into kinetic viscosity and collisional viscosity [Bakker, 2008].

$$\mu_s^{kc} = \mu_{s,\text{kin}} + \mu_{s,\text{coll}} \tag{2.33}$$

It is calculated with a modified version of Hrenya and Sinclair [1997] and Agrawal et al. [2001] that accounts for the influence of boundaries, and is given by Schneiderbauer et al. [2012a] with

$$\mu_s = \frac{18}{15} \left\{ \frac{\mu}{g_0 \eta_s \left(2 - \eta\right)} \left(\frac{1}{1 + \frac{l_s}{L_C}} + \frac{8}{5}\alpha_s \eta_s g_0 \right) \left(1 + \frac{8}{5}\eta_s \left(3\eta_s - 2\right)\alpha_s g_0 \right) + \frac{3}{5}\eta_s \mu_b \right\}, \tag{2.34}$$

where

$$\mu = \frac{5\rho_s d_s \sqrt{\pi\Theta}}{96}, \tag{2.35}$$

which is proportional to the particle collisions frequency ($f \propto \sqrt{\Theta}$) and

$$\mu_b = \frac{256\mu\alpha_s^2 g_0}{5\pi}. \tag{2.36}$$

The bulk viscosity accounts for the resistance of the particle ensemble to compression and expansion. It is given by

$$\lambda_s^{kc} = \frac{8}{3} \eta_s \alpha_s^2 \rho_s d_s g_0 \sqrt{\frac{\Theta}{\pi}}. \tag{2.37}$$

In the Eulerian granular model, a thermodynamic collisional pressure for the solid phase is defined such that it is equivalent to the gas pressure. It is described by the model of Hrenya and Sinclair [1997] and Lun et al. [1984]. It consists of a kinetic term and a term due to particle collisions and is given by

$$p_s^{kc} = \alpha_s \rho_s \left(\frac{1}{1 + \frac{l_s}{L_C}} + 4\eta_s \alpha_s g_0 \right) \Theta. \tag{2.38}$$

2.6.2 Frictional stresses

In the frictional regime (volume fraction $\alpha_s > 0.5$) the particle-particle collisions are no longer instantaneous as assumed by kinetic theory. In this regime there are long sliding frictional contacts which does not apply in the kinetic theory. Therefore, the frictional part of the solids shear viscosity does not depend on the amount of pseudo-thermal energy [Srivastava and Sundaresan, 2003b, Schneiderbauer et al., 2012a]. It is common to model the yield stress, at which the granular material begins to flow, by Coulomb's law [Schaeffer, 1987, van Wachem et al., 2001]

$$\tau_s^{fr} = \mu_i p_s^{fr}, \tag{2.39}$$

where τ_s^{fr} denotes the frictional shear stress, p_s^{fr} the frictional pressure and $\mu_i = \sin \phi_i$ is the coefficient of internal friction with ϕ_i the angle of the internal friction. The frictional stresses are usually written in non-Newtonian form [Schneiderbauer et al., 2012a, Schaeffer, 1987]. The rigid-plastic rheological model for the frictional stresses is given by [Srivastava and Sundaresan, 2003b]

$$\mathbf{S}_s^{fr} = p_s^{fr} \mathbf{I} - 2\mu_s^{fr} \left(\mu_i, p_s^{fr}, \text{dev}\mathbf{D}_s \right) \text{dev}\mathbf{D}_s, \tag{2.40}$$

where $\text{dev}\mathbf{D}_s$ denotes the deviatoric part of $\mathbf{D}_s$ (2.32). By assuming Coulomb friction, Schaeffer [1987] derived the frictional viscosity from first principles to

$$\mu_s^{fr} = \frac{\mu_i p_s^{fr}}{2\|\text{dev}\mathbf{D}_s\|}. \tag{2.41}$$

23

with $\|\text{dev}\mathbf{D}_s\| = \sqrt{\mathbf{D}_s : \mathbf{D}_s / 2}$. Note that μ_s^{fr} diverges by numerical computation. Various formulations have been proposed in literature for the frictional pressure. For example Johnson and Jackson [1987], Johnson et al. [1990] and Jackson [2000] suggested

$$p_s^{\text{fr}} = F \frac{(\alpha_s - \alpha_{s,\min})^r}{(\alpha_{s,\max} - \alpha_s)^p}, \tag{2.42}$$

where F, r and p are material constants. For example, $F = 0.05$, $r = 2$ and $p = 5$ are typically used for glass beads. $\alpha_{s,\min}$ denotes the value of volume fraction at which frictional interaction occurs, which is typically around $\alpha_{s,\min} = 0.5$. Also Syamlal [1987] proposed a simple empirical function for the frictional pressure,

$$p_s^{\text{fr}} = A \left(\alpha_s - \alpha_{s,\min}\right)^n, \tag{2.43}$$

where $A = 10^{25}$ and $n = 10$ are material constants and common values used for the simulation of frictional pressure in a mixture of glass beads. The two presented functions for frictional pressure are monotonically increasing with increasing α_s [Srivastava and Sundaresan, 2003a]. Schneiderbauer et al. [2012a] and Schneiderbauer and Pirker [2012b] suggested a model that recognizes the shear rate dependent rheology and dilatation in the frictional regime [da Cruz et al., 2005, Chialvo et al., 2012]. The frictional pressure reads

$$p_s^{\text{fr}} = 4\rho_s \left(\frac{bd_s \|\text{dev}\mathbf{D}_s\|}{\alpha_{s,\max} - \alpha_s}\right)^2 \tag{2.44}$$

where $b = 0.2$ [Forterre and Pouliquen, 2008] for mono-dispersed glass beads. The frictional viscosity is given by

$$\mu_s^{\text{fr}} \left(I_s, p_s^{\text{fr}}, \text{dev}\mathbf{D}_s\right) = \frac{\mu_i (I_s) p_s^{\text{fr}}}{2\|\text{dev}\mathbf{D}_s\|} \tag{2.45}$$

where

$$I_s = \frac{2\|\text{dev}\mathbf{D}_s\|}{\sqrt{\frac{p_s^{\text{fr}}}{\rho_s}}} \tag{2.46}$$

and

$$\mu_i = \mu_i^{\text{st}} + \frac{\mu_i^c - \mu_i^{\text{st}}}{\frac{I_0}{I_s} + 1} \tag{2.47}$$

24

obtained by Jop et al. [2006], where μ_i^c, μ_i^{st} and I_0 are constants. Typical values for the constants, for mono-dispersed glass beads are $I_0 = 0.279$, $\mu_i^{st} = \tan{(20.9^\circ)}$ and $\mu_i^c = \tan{(32.76^\circ)}$ [Forterre and Pouliquen, 2008].

2.7 Turbulence modelling

In the preheating tower, the gas flow has a Reynolds number of more than Re = 10^5. Furthermore, the gas flow is turbulent. One problem due to computational limitations, is that the grid size must be in a range where it is not possible to resolve all turbulent scales. Hence, the turbulence of the gas phase is commonly represented by turbulence models to account for the influence of unresolved eddies in the mean flow. Turbulence models that are commonly used in industrial cases are the Reynolds Averaged Navier-Stokes (RANS) models (for example $k\epsilon$, RSM) because they include the influence of small vortices. In literature there are two basic approaches to extend single phase RANS to multiphase flows. One approach is the mixture approach and the second one is the dispersed approach. In the mixture approach a dilute to medium density particulate flow is assumed and for the the dispersed approach a dilute particulate flow is assumed. Furthermore, in dense particulate flow regimes, the gas flow through the particles is assumed to be laminar. The mixture model is a good choice if the density ratio between the phases is around one [ANSYS, 2009], which is not the case in our flow situation. Consequently, in this work we used the dispersed RANS model as turbulence model, because it is important to model the physics of turbulence in dispersed regions correctly.

2.8 Boundary Conditions

The main challenge at the wall is to calculate the shear stresses τ_s^{kc} and the flux of fluctuation energy, **q**, from the wall into the domain. Particles hitting a wall can either slide, roll or are directly reflected back. Johnson and Jackson [1987] presented a model, which is nowadays commonly used for simulations. However, this model ignores the fact that a granular medium sliding at a wall can only exert a shear stress limited by Coulumbs law to the wall. This, in term, implies that these boundary conditions overestimate τ_s^{kc} and q in rapid granular flows. This is recognized by the model of Jenkins and Louge [1997]. The paper of Schneiderbauer et al. [2012b] is a generalization of the boundary conditions derived in Jenkins and Louge [1997]. The new set of boundary conditions for the Eulerian phase described in Schneiderbauer et al. [2012b] has been applied during this survey. In the case of the preheating tower, simulating wall bounded particle conveying is crucial because of the high impact of walls on strand formated particle-laden flows. Obviously, this requires a thorough modelling of particle-wall interactions. Hence this new boundary condition should be used for the simulations.

2.8.1 Johnson and Jackson

Johnson and Jackson [1987] presented a model to calculate the shear stress of particles at bounding walls, based on an earlier model given by Hui et al. [1984]. The model for the particle flux, $\dot{m}_\mathrm{p}$, to the wall consists of three main factors: (i) the collision frequency, (ii) the particle mass and (iii) the affected boundary area. The collision frequency, f_coll, between particles and wall is given by

$$f_\mathrm{coll} = \frac{\sqrt{3\Theta}}{x} \tag{2.48}$$

where x denotes the averaged particle distance to the wall and can be assumed to be $x = s/2$, which is half of the average distance between the particles. The mass of a particle is given by

$$m_\mathrm{p} = \frac{d_\mathrm{p}^3 \pi}{6} \rho_\mathrm{p}. \tag{2.49}$$

The affected area, a_c, per particle is defined by l^2 and using Equation (2.20) it can be written as

$$a_c = l^2 = d_\mathrm{p}^2 \left(\frac{\alpha_\mathrm{s,max}}{\alpha_\mathrm{s}} \right)^{\frac{2}{3}}. \tag{2.50}$$

With these three factors the particle flux to the wall can be calculated. It is given by [Johnson and Jackson, 1987]

$$\dot{m}_\mathrm{p} = \frac{\sqrt{3\Theta}}{x} \frac{d_\mathrm{p}^3 \pi}{6} \rho_\mathrm{p} \frac{1}{d_\mathrm{p}^3 \left(\frac{\alpha_\mathrm{s,max}}{\alpha_\mathrm{s}} \right)^{\frac{2}{3}}}, \tag{2.51}$$

which can be simplified to

$$\dot{m}_p = \frac{\pi}{6}\rho_s\sqrt{3\Theta}\frac{\alpha_s}{\alpha_{s,\text{max}}}g_0. \qquad (2.52)$$

Since the normal velocity to the wall must be zero, the shear stress to the wall can finally be written as [Schneiderbauer et al., 2012a]

$$\tau_s^{\text{kc}} = -\frac{\sqrt{3\pi}}{6}\phi'\frac{\alpha_s\rho_s g_0}{\alpha_{s,\text{max}}}\sqrt{\Theta}\mathbf{u}_s^{\text{sl}} \qquad (2.53)$$

where ϕ' denotes the specularity coefficient [Hui et al., 1984] and $\mathbf{u}_s^{\text{sl}}$ is the slip velocity at the wall. The specularity coefficient was introduced by Hui et al. [1984] to quantify the nature of particle-wall collisions and used by Johnson and Jackson [1987]. It is an empirical global constant ranging between 0 and 1, and represents the impact of wall parameters (e.g. wall roughness). For $\phi' = 0$, a smooth frictionless wall is present and therefore no shear stress is exerted by the wall. For $\phi' = 1$, a very rough wall is present and the complete slip velocity is translated into shear stress. Typical values for the specularity coefficient are between 0.2 and 0.6 depending on the roughness and curvature of the wall. The flux of fluctuation energy at the bounding walls, needed for the boundary condition of the granular temperature, is given by

$$\mathbf{n}\cdot\mathbf{q} = \tau_s^{\text{kc}}\cdot\mathbf{u}_s^{\text{sl}} + \frac{\sqrt{3\pi}}{4}\frac{\alpha_s\rho_s g_0}{\alpha_{s,\text{max}}}\left(1 - e^2\right)\Theta^{\frac{3}{2}}. \qquad (2.54)$$

Although the boundary conditions of Johnson and Jackson [1987] had a huge impact on the CFD simulation community, there is still some uncertainty. The shear stress to the wall increases linearly with the slip velocity $\mathbf{u}_s^{\text{sl}}$, while in reality there is a threshold given by Coulombs law. Furthermore, the specularity coefficient should not be a global constant. Rather it should depend on the local flow configuration at the particle-wall collision position.

2.8.2 Li and Benyahia

The specularity coefficient, introduced by Hui et al. [1984] and used by Johnson and Jackson [1987], depends on the curvature, roughness and restitution coefficient of the wall. Recently, Li and Benyahia [2012] revised the boundary conditions of Johnson and Jackson and expressed the specularity coefficient as a function of particle-wall restitution coefficient, frictional coefficient and the normalized slip velocity $\|\mathbf{u}_s^{sl}\|$ at the wall. They suggested

$$
\phi' = \left\{
\begin{array}{ll}
-\frac{7\sqrt{6\pi}\left(\phi_0'\right)^2}{8\frac{7}{2}\mu_w(1+e_w)} \frac{\|\mathbf{u}_s^{sl}\|}{\sqrt{3\Theta}} + \phi_0' & \|\mathbf{u}_s^{sl}\| \leq \frac{2\mu_w(1+e_w)\sqrt{3\Theta}}{\sqrt{6\pi}\phi_0'} \\
\frac{\mu_w(1+e_w)\sqrt{\Theta}}{\|\mathbf{u}_s^{sl}\|} & \text{otherwise}
\end{array}
\right.
\tag{2.55}
$$

where ϕ_0' denotes the value of ϕ' when $\|\mathbf{u}_s^{sl}\|$ is zero and e_w is the restitution coefficient between particle and wall. With this model the specularity coefficient can be described with physical parameters and guarantees that the shear stress is limited by Coulomb's friction. Thus, this is an improvement to the boundary conditions of Johnson and Jackson, and includes the effect of particle sliding and rolling at the wall during the collision phase.

2.8.3 Jenkins and Louge

Jenkins [1992] derived boundary conditions for a flat, frictional wall based on a collision model that distinguishes between sliding and sticking particle-wall collisions. The shear stress in the "large friction/no sliding" limit is given by

$$
\tau_s^{kc} = N\frac{3}{7}\frac{1+\beta_0}{1+e_w}\frac{\|\mathbf{u}_s^{sl}\|}{\sqrt{3\Theta}},
\tag{2.56}
$$

where N denotes the normal stress to the wall given by

$$
N = \eta_w\alpha_s\rho_s g_0\Theta,
\tag{2.57}
$$

with

$$
\eta_w = 0.5\left(1 + e_w\right).
\tag{2.58}
$$

For the "small friction/all sliding" limit, τ_s^{kc} is given by

$$
\tau_s^{kc} = N\mu_w
\tag{2.59}
$$

where μ_w denotes the coefficient of wall friction. A few years later, Jenkins and Louge [1997] improved these boundary conditions by refining the calculation of the

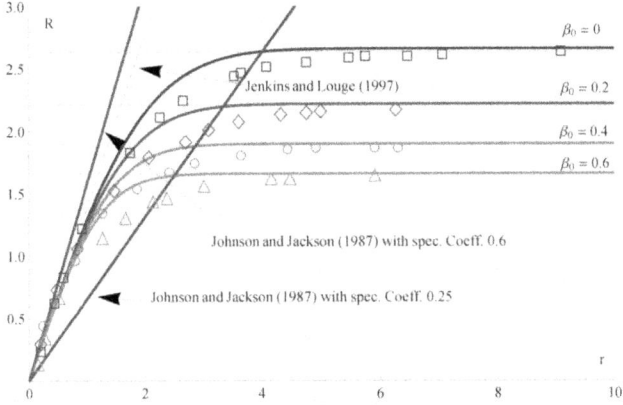

Figure 2.5: Schneiderbauer et al. [2012b] boundary conditions: Stress ratio $R = \frac{7}{2}\frac{1+e_w}{1+\beta_0}\frac{S}{N}$ for different tangential restitution coefficients β_0 over normalized slip $r = \frac{u_{sl}}{\sqrt{3\Theta}}$. The oblique and horizontal dotted lines represent Jenkins no sliding and all sliding limits. The $\circ$, $\triangle$, $\square$ and $\Diamond$ symbols indicate results from simulations done by Louge [1994].

flux of fluctuation energy between particles and wall. Jenkins and Louge [1997] found that the maximum energy flux, for the sliding case, is

$$\mathbf{q} = N\sqrt{3\Theta}\frac{2}{1+e_w}\sqrt{\frac{2}{3\pi}}\left[\frac{1}{7}\mu_0^2 - \frac{1}{2}\left(1 - e_w^2\right) - \mu_0\mu_w e_w\left(\frac{1+e}{e+2/e_s}\right)\right], \qquad (2.60)$$

where e_s is the restitution factor between the particles, β_0 is the tangential coefficient of restitution and

$$\mu_0 = \frac{7}{2}\frac{1+e_w}{1+\beta_0}\mu_w. \qquad (2.61)$$

The energy flux before reaching the maximum is given by

$$\mathbf{q} = -N\sqrt{3\Theta}\sqrt{\frac{\pi}{6}}\left\{\frac{2}{\pi}\left(1 - e_w\right) + \frac{2}{7}\frac{1-\beta_0^2}{1+e_w}\right.$$

$$\left[\sin^4\phi_0 + r^2\sin^2\phi_0\left(2\sin^4\phi_0 - 2\sin^2\phi_0\cos^2\phi_0 - \frac{4}{1-\beta_0}\sin^2\phi_0\right)\right]$$

$$+\frac{\mu_w}{2}\left[\pi - 2\sin\phi_0\cos\phi_0 4\sin\phi_0\cos^3\phi_0 - 2\phi_0 + r^2\left(-\pi + 2\sin\phi_0\cos\phi_0 2\phi_0 +\right.\right.$$

$$\left.16\sin^3\phi_0\cos^3\phi_0 - 4\sin\phi_0\cos^3\phi_0\right)\right] - \mu_w\mu_0\left[1 - 2\sin^2\phi_0\sin^4\phi_0 +\right.$$

$$\left.\left.r^2\left(2\sin^2\phi_0\cos^2\phi_0 + sin^2\phi_0\cos^4\phi_0 - 2\sin^4\phi_0\cos^2\phi_0\right)\right]\right\}, \qquad (2.62)$$

29

where

$$\tan \phi_0 = \frac{\mu_0}{1 + \beta_0},$$ (2.63)

and

$$r \equiv \frac{\|\mathbf{u}_s^{sl}\|}{\sqrt{3\Theta}}$$ (2.64)

denotes the ratio of the mean velocity of the contact point and the boundary. For bumpy, frictional walls with cylindrical and spherical bumps, Jenkins [2001] outlined additional boundary conditions for granular flows of frictional spheres. For more information, a detailed comparison of the results from Jenkins and Louge [1997] and computer simulations done by Louge [1994] can be found in Schneiderbauer et al. [2012b].

2.8.4 Schneiderbauer et. al.

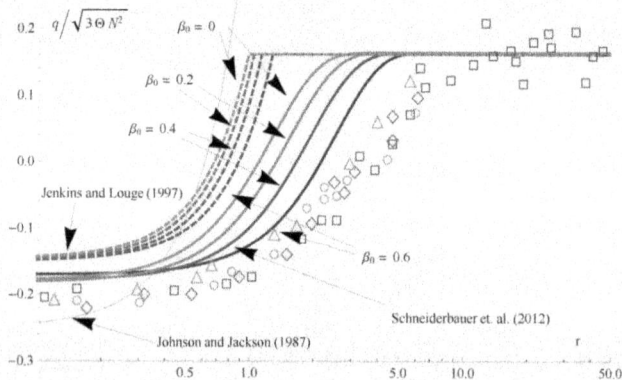

Figure 2.6: Normalized flux of fluctuation energy, $q/\left(\sqrt{3\Theta}N\right)$ of the three models, Johnson and Jackson [1987], Jenkins and Louge [1997] and Schneiderbauer et al. [2012b] over normalized slip $r = \frac{u_{sl}}{\sqrt{3\Theta}}$.

Based on the boundary conditions of Jenkins [1992], Schneiderbauer et al. [2012b] developed new boundary conditions for rapid granular flows. The wall shear stresses are given by Schneiderbauer et al. [2012a] and Schneiderbauer and Pirker [2012b] as,

$$\tau_s^{kc} = -\eta_w \mu_w \alpha_s \rho_s g_0 \Theta \mathrm{erf}\left(\bar{u}_s^{sl}\right) \frac{\mathbf{u}_s^{sl}}{\|\mathbf{u}_s^{sl}\|}, \tag{2.65}$$

including

$$\bar{u}_s^{sl} = \frac{\mathbf{u}_s^{sl}}{\sqrt{2\Theta}\mu_0}. \tag{2.66}$$

The flux of fluctuation energy at the bounding wall is given by

$$\mathbf{n} \cdot \mathbf{q} = \tau_s^{kc} \cdot \mathbf{u}_s^{sl} - \frac{\alpha_s \rho_s g_0 \eta_w \sqrt{\Theta}}{\sqrt{2}\mu_0^2 \sqrt{\pi}} \exp\left(-\bar{u}_s^{sl^2}\right) \left\{ \mu_w \left[2\mu_w \|\mathbf{u}_s^{sl}\|^2 (2\eta_w - \mu_0) + \right.\right.$$
$$\Theta \left(14\mu_w \eta_w - 4\mu_0 (1 - \mu_w) - 6\mu_w \mu_0^2 \eta_w\right) \right]$$
$$+\mu_0^2 \sqrt{\Theta} \exp\left(\bar{u}_s^{sl^2}\right) \left[\sqrt{\Theta} \left(4 (\eta_w - 1) + 6\mu_w^2 \eta_w\right) - \sqrt{2\pi}\mu_w \|\mathbf{u}_s^{sl}\| \mathrm{erf}\left(\bar{u}_s^{sl}\right) \right] \right\} \tag{2.67}$$

In Figure 2.5, the stress ratio, R, for different tangential coefficients of restitution, β_0, is shown. The oblique and horizontal dotted lines represent Jenkins' no sliding and all sliding limits. It can be observed that the model of Schneiderbauer et al. [2012b] reaches the limits given by Jenkins and Louge [1997] when $\frac{u_{slip}}{\sqrt{3\Theta}} = 0\,\mathrm{m/s}$ and $\frac{u_{slip}}{\sqrt{3\Theta}} > 4$. In between a smooth transition from one limit to the other is realized. This agrees with the simulations done by Louge [Schneiderbauer et al., 2012b]. In Fig. 2.6 the normalized flux of fluctuation energy for different friction coefficients is shown. It can be seen that with increasing friction coefficient the flux q is increasing. Furthermore, this boundary conditions also account for higher energy loss at a rough wall between wall and granular material as at flat walls. This indicates that particles in the granular material are decelerated more by a rough wall than by a flat one. Consequently, this effect increases the pressure loss in a particle conveying system. In the case of the preheating tower, good modelling of very rough walls (lined walls) is important. Therefore, this wall boundary condition is recommended for the simulations.

3

Lagrangian discrete phase modelling

In dilute particle-laden flows particle-particle collisions are very infrequent and the particle movement depends mostly on the interaction of fluid and particles. Furthermore, the number of particles is computational manageable to handle each particle on its own. Hence, a model which depicts the movement of single particles can be used. As already mentioned in the introduction, the Lagrangian discrete phase model (DPM) can be taken for the simulation of single particles, or a particle cloud. This model is commonly used in dilute to medium density particle-laden flows (Figure. 3.1). Particle movement is calculated using Newton's second law for each particle.

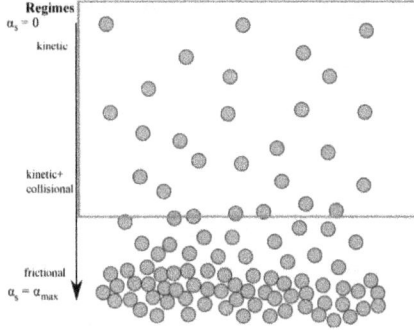

Figure 3.1: Range of validity of the discrete phase model

3.1 Force balance and torque balance

The acceleration, $\frac{d}{dt}\mathbf{u}_p$, of the particle is given by Newton's second law (Figure 3.2).

$$\frac{d}{dt}\mathbf{u}_p = \sum \mathbf{f}_p, \qquad (3.1)$$

where $\mathbf{f}_p = \mathbf{F}/m_p$ denotes the force per mass on a particle. By considering the most important forces, which are the drag force and the gravitational force, the above equation can be rewritten to

$$\frac{d}{dt}\mathbf{u}_p = \mathbf{f}_{p,drag} + \mathbf{g} + \mathbf{f}_{p,add}, \qquad (3.2)$$

where $\mathbf{f}_{p,add}$ denotes additional forces, e.g. Magnus force, Saffman force and particle-particle interaction forces.

Similarly to the translational force balance, a rotational momentum balance can be defined

$$\frac{d}{dt}\boldsymbol{\omega}_p = \mathbf{t}_{p,g} + \mathbf{t}_{p,add}, \qquad (3.3)$$

where $\mathbf{t}_{p,g}$ denotes the torque on a particle exerted from the fluid phase and $\mathbf{t}_{p,add}$ denotes additional torque sources arising, as for example, from particle-particle collisions [Kahrimanovic, 2009].

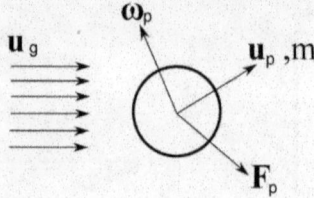

Figure 3.2: Forces and parameters of a discrete phase particle

3.2 Forces on a particle

In this section we give a short overview on the important and dominant forces on particles for pneumatic conveying and separation in the preheating tower. For a more detailed introduction and review of the forces acting on particles the reader is referred to Crowe et al. [1998] and Kahrimanovic [2009], and the literature cited therein.

3.2.1 Drag force

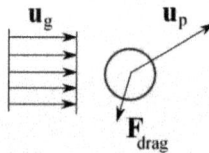

Figure 3.3: Drag force on the particle

In the case of pneumatic conveying, the most dominant force between particles and surrounding fluid is the drag force. It is based on the velocity difference between the center of mass of the particle and the fluid. The force is always in the direction of the velocity difference of fluid and particle velocity. It can be expressed as follows

$$f_{p,drag} = \beta_p \left(u_g - u_p \right),$$

(3.4)

34

where $\mathbf{u}_g$ denotes the gas velocity and $\mathbf{u}_p$ the particle velocity. In this case, the interphase drag coefficient β_p is calculated with the model of Wen and Yu [1966]

$$\beta_p = \alpha_g \frac{18\mu_g}{\rho_p d_p^2} \frac{C_D \mathrm{Re}_p}{24}. \tag{3.5}$$

α_g denotes the volume fraction of the fluid, μ_g describes the fluid viscosity, ρ_p is the density of the particle, d_p the diameter of the particle, and C_D is the drag coefficient. Note, as mentioned in Chapter 2, also other models are available for the interphase momentum exchange. The particle Reynolds number Re_p is given by

$$\mathrm{Re}_p = \frac{\rho_g d_p |\mathbf{u}_g - \mathbf{u}_p|}{\mu_g}. \tag{3.6}$$

The model of Schiller and Naumann [1935] is commonly used for the drag coefficient. It is given by

$$C_D = \begin{cases} \dfrac{24\left(1+0.15(\alpha_g \mathrm{Re}_p)^{0.687}\right)}{\alpha_g \mathrm{Re}_p} & \mathrm{Re}_p \leqslant 1000 \\[2ex] 0.44 & \mathrm{Re}_p > 1000 \end{cases} \tag{3.7}$$

Additional models for the drag coefficient can be found in Crowe et al. [1998].

3.2.2 Particle rotation and Magnus force

By visiting tennis games Newton [1671-72] observed that a rotating ball flies in a curve. Nowadays it is generally known that every rotating ball or particle flies in a curve normal to the rotation axis. Magnus [1852] was the first to describe this force mathematically. The Magnus force depends on the particle fluid rotation velocity, Ω_p [Magnus, 1852, Barkla, 1971], which describes the difference between particle velocity and fluid velocity on the particle surface. It is given by

$$\Omega_p = \frac{1}{2}\nabla \times \mathbf{u}_g - \boldsymbol{\omega}_p. \tag{3.8}$$

where $\boldsymbol{\omega}_p$ describes the particle angular velocity. The Magnus force (Figure 3.4) is given by

$$\mathbf{f}_{p,\mathrm{Magnus}} = \frac{1}{2}|\mathbf{u}_g - \mathbf{u}_p| C_L \left(\frac{d_p}{2}\right)^2 \pi \frac{\Omega_p \times (\mathbf{u}_g - \mathbf{u}_s)}{|\Omega_p|} \frac{\rho_g}{m_p} \tag{3.9}$$

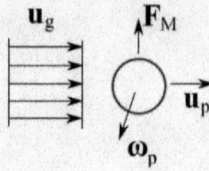

Figure 3.4: Magnus force

where C_L denotes the rotating coefficient and m_p the particle mass. In particle-laden flows with dilute loading and only few particle-particle collisions, the Magnus force can not be ignored. In dense particle-laden flows, where the main interaction comes from particle-particle collisions and frictional contributions, the Magnus force is negligible because the rotations of the particles are damped by collision. In computational fluid dynamics, dimensionless numbers are often used to describe a force. For the Magnus force, this can be done by the rotational Reynolds number, given by [Kahrimanovic, 2009]

$$\mathrm{Re_R} = \frac{\rho_g d_p^2 |\mathbf{\Omega_p}|}{\mu_g}. \tag{3.10}$$

3.2.3 Saffman force

Saffman [1965] studied particles in a shear flow (Figure 3.5). He observed a force acting on particles depending on the velocity gradient of the continuous phase around the particle. Furthermore, Saffman [1965] analyzed and described the resulting

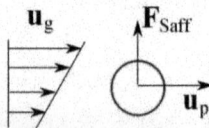

Figure 3.5: Saffman force

buoyancy force for small Reynolds numbers and found an analytic solution [Kahri-manovic, 2009]. It is given by

$$\mathbf{F}_{\text{Saff}} = 1.615\mu_g \mathbf{u} d_p^2 \sqrt{\frac{1}{\nu_g}\beta}, \tag{3.11}$$

where β denotes the shear rate. In one dimensional description it is written as

$$\beta = \frac{du}{dy}. \tag{3.12}$$

Introducing the particle Reynolds number for the deviation of the velocities with

$$\text{Re}_\beta = \frac{d_p^2}{\nu_g}\beta, \tag{3.13}$$

changes to Equation 3.11

$$\mathbf{F}_{\text{Saff}} = 1.615\mu_g \bar{u} d_p \sqrt{\text{Re}_\beta}. \tag{3.14}$$

The validity for the equation is given by Saffman [1965] for

$$\left.\begin{array}{c} \text{Re}_p \\ \text{Re}_R \\ \text{Re}_\beta \end{array}\right\} \ll 1 \tag{3.15}$$

and

$$\text{Re}_p \ll \sqrt{\text{Re}_\beta}. \tag{3.16}$$

The particle Reynolds number in pneumatic conveying, as in the preheating tower, is often $\text{Re}_p > 1$. From experiments and simulations enhancements of existing models for C_{LS} were deduced by Dandy and Dwyer [1990], McLaughlin [1991], Mei [1992] and Crowe et al. [1998] to create a model for higher Reynolds numbers. This yields the Saffman force equation given by Crowe et al. [1998] with

$$\mathbf{F}_{\text{Saff}} = 1.61 d_p^2 \frac{\sqrt{\mu_g \rho_g}}{\sqrt{|\nabla \times \mathbf{u}_g|}} \left[(\mathbf{u}_g - \mathbf{u}_p) \times (\nabla \times \mathbf{u}_g)\right]$$

$$\begin{cases} \left(1 - 0.234\sqrt{\beta'}\right)e^{-\frac{\text{Re}_p}{10}} + 0.234\sqrt{\beta'} & \dots \quad \text{Re}_p \leq 40 \\ 0.0371\sqrt{\beta'\text{Re}_p} & \dots \quad \text{Re}_p > 40 \end{cases}. \tag{3.17}$$

In literature, the Saffman force is commonly transformed into a dimensionless buoyancy force C_{LS} given by

$$C_{LS} = \frac{|\mathbf{F}_{\text{Saff}}|}{\frac{1}{2}\rho_g \bar{\mathbf{u}}^2 \frac{d_p^2 \pi}{4}} = 4.11 \sqrt{\frac{\beta'}{\text{Re}_p}} \tag{3.18}$$

where β' denotes the dimensionless shear rate

$$\beta' = \beta \frac{d_p}{\bar{u}}. \tag{3.19}$$

3.2.4 Additional forces

A pressure gradient introduces a force which accelerates the air and particles [Stull, 2000, Crowe et al., 1998]. It is given by

$$\mathbf{F}_p = -\frac{m_g}{\rho_g} \nabla p \tag{3.20}$$

where m_g denotes the displaced mass of gas. Furthermore, with the acceleration of a particle, a virtual mass force arises. This force describes the acceleration of the fluid that surrounds the particle, which can be described as a "virtual mass" that is added to the particle mass. It is given by [Crowe et al., 1998]

$$\mathbf{F}_{\text{VM}} = c_A m_g \frac{\mathrm{d}}{\mathrm{d}t} \left(\mathbf{u}_g - \mathbf{u}_p \right), \tag{3.21}$$

where c_A describes a factor which is about 0.5. Finally, the Basset force, which was first mentioned by Basset [1881] and Boussinesq [1903], accounts for viscous effects and is given by [Crowe et al., 1998]

$$\mathbf{F}_{\text{Basset}} = 9\sqrt{\frac{\rho_g \mu_g}{\pi}} \frac{m_p}{\rho_p d_p} c_H \int_{-\infty}^{t} \frac{\frac{\mathrm{d}}{\mathrm{d}\tau}\left(\mathbf{u}_g - \mathbf{u}_p\right)}{\sqrt{(t - \tau)}} \mathrm{d}\tau \tag{3.22}$$

In practical situations the Basset force is often neglected, but for cases with high acceleration rate, it should also be considered. Furthermore, $\mathbf{F}_p$ and $\mathbf{F}_{\text{VM}}$ can be neglected here since $\frac{\rho_g}{\rho_p} \ll 1$.

3.3 Torque

Every rotating particle is interacting with the surrounding fluid. Rotation energy is transported via surface contacts between the particle and the gas. Hence angular momentum, which is described by $t_{p,g}$ in equation (3.3) is lost or gained. $t_{p,g}$ is given by [Kahrimanovic, 2009]

$$t_{p,g} = \frac{\mu_g}{64 I_p} d_p^3 C_R \Omega_p Re_R. \tag{3.23}$$

where I_p describes the inertia of a spherical particle

$$I_p = \frac{m_p d_p^2}{10} \tag{3.24}$$

and Re_R is the particle rotational Reynolds number

$$Re_R = \frac{\rho_g d_p^2 |\Omega_p|}{\mu_g}. \tag{3.25}$$

The drag coefficient of rotation is calculated with

$$C_R = \begin{cases} 64 \frac{\pi}{Re_R} & Re_R \leqslant 32 \\ \frac{12.9}{\sqrt{Re_R}} + \frac{128.4}{Re_R} & 32 < Re_R < 1000 \end{cases} \tag{3.26}$$

A more detailed discussion of the torque can be found in Sawatzki [1970], Dennis et al. [1980], Sommerfeld [2000], Kahrimanovic [2009] and the literature cited therein.

3.4 Turbulent fluctuations

In general, the trajectory of a particle moving through turbulent flow is influenced by the turbulent eddies. Hence the influence of the velocity fluctuations on the drag has to be considered. The impact of turbulent eddies on the particle movement increases with decreasing Stokes number, which is given by

$$St_p = \frac{\tau_p}{\tau_e} \tag{3.27}$$

where τ_p denotes the particle relaxation time given by

$$\tau_p = \frac{\rho_p d_p^2}{18\mu_g} \left(1 + 0.15 \mathrm{Re}_p^{0.687}\right)^{-1} \tag{3.28}$$

and τ_e is the eddy lifetime. One model for including the velocity fluctuations into the particle track is the Random Walk Model (RWM) [Gosman and Ioannides, 1983]. The RWM includes the effect of instantaneous turbulent velocity fluctuations on the particle trajectories through the use of stochastic methods [ANSYS, 2009]. This is done by dividing the velocity into a mean and a fluctuating part given by

$$\mathbf{u}_p = \bar{\mathbf{u}}_p + \mathbf{u}_p', \tag{3.29}$$

where $\bar{\mathbf{u}}_p$ denotes the average velocity and $\mathbf{u}_p'$ the stochastic fluctuating part of the velocity. $\mathbf{u}_p'$ is computed for n_{RWM} representative particles at each trajectory to get representative effects of the turbulent particulate flow. If $\mathbf{u}_p'$ changes value after a fixed number of time steps, then the model is called discrete random walk model (DRW) or "eddy-lifetime" model [ANSYS, 2009]. In this commonly used model, the interaction of particles and fluid phase eddies is characterized by the Gaussian-distributed random velocities fluctuations, u', v', and w' and the eddy lifetime τ_e

$$\tau_e = -T_L \log(r) \tag{3.30}$$

where r is a uniformly distributed random number between 0 and 1. For Reynolds Average Navier Stokes (RANS) turbulence models, T_L is given by

$$T_L \approx \zeta \frac{k}{\epsilon} \tag{3.31}$$

where ζ is a normally distributed random number and k and ϵ are parameters of the turbulent gas flow field. The random Gaussian-distributed velocity fluctuations are given by

$$u' = \zeta \sqrt{\overline{u'^2}} \tag{3.32}$$

where $\overline{u'^2}$ describes the root mean square value of the fluctuation velocity. Assuming isotropy and knowing the turbulent flow field with its parameters k, ω or ϵ results in

$$\sqrt{\overline{u'^2}} = \sqrt{\overline{v'^2}} = \sqrt{\overline{w'^2}} = \sqrt{\frac{2k}{3}}. \tag{3.33}$$

If the Reynolds Stress Model (RSM) model is used for turbulent models, non-isotropy is included and the velocity fluctuations are separately given by [ANSYS, 2009]

$$
\begin{aligned}
u' &= \zeta\sqrt{\overline{u'^2}} \\
v' &= \zeta\sqrt{\overline{v'^2}} \\
w' &= \zeta\sqrt{\overline{w'^2}}.
\end{aligned}
\tag{3.34}
$$

Hence, $\mathbf{u}_p$ is used in the computation of the particle track, which now includes velocity fluctuations. Obviously, every particle track starting at the same point will end up with another trajectory, if the random number generator provides different random numbers.

3.5 Particle wall collisions

In industrial conveying processes, e. g. the preheating tower, it is observed that the impact of the wall on the particle trajectories is important. Particles colliding with the wall are decelerated. Hence, the particles are continuously accelerated by the fluid implying a deceleration of the flow. Consequently, this leads to a higher pressure loss. For very small particles in dilute flows, the wall collisions are not relevant for the pressure loss because these particles have a very low inertia. Such small particles promptly reach the speed of the surrounding fluid after a wall collision. Bigger particles with higher inertia times must be accelerated for a longer time. Hence, these particles are mainly responsible for the pressure loss in a pneumatic conveying system. Additionally, it is observed that with increasing wall roughness, the pressure loss increases too. To account for this effect, we use models that are based on the wall parameters, the particle values, impact angle and the restitution- and friction coefficients between particles and wall. Firstly, a simple model for smooth, flat walls is presented. This model is based on restitution coefficients for both normal and tangential directions. Secondly, a model for rough walls is given [Tsuji et al., 1989, Sommerfeld and Huber, 1999, Sommerfeld, 2001].

3.5.1 Restitution coefficient model

This model assumes an ideal, smooth wall. The momentum loss during a particle-wall collision is therefore described as ratio of the speed before and after the impact.

Furthermore, it is divided into normal and tangential parts which are characterized by e_n and e_t. The velocities in normal and tangential direction, after reflection, can be computed from

$$\mathbf{u}_{n,after} = e_n \mathbf{u}_{n,before} \tag{3.35}$$

and

$$\mathbf{u}_{t,after} = e_t \mathbf{u}_{t,before}, \tag{3.36}$$

where usually e_t is set to $\frac{e_n}{2}$. Common values for glass beads or sand are $e_t = 0.4$

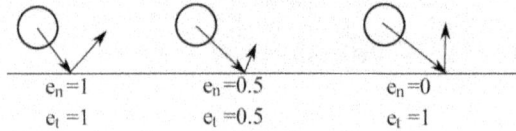

| $e_n = 1$ | $e_n = 0.5$ | $e_n = 0$ |
| $e_t = 1$ | $e_t = 0.5$ | $e_t = 1$ |

Figure 3.6: Example for different restitution coefficients

and $e_n = 0.9$ and are used in this study. If only smooth walls are present, the restitution coefficient model is the most computationally efficient model that yields appropriate results. However, for the case of the preheating tower, no smooth walls are present, so a more physical model that includes rough walls is required.

3.5.2 Rough wall-particle collisions

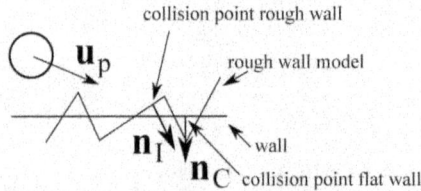

Figure 3.7: Particle wall collision with rough wall. The difference of the normal vector between flat wall n_C and the rough wall n_I is shown.

In Figure 3.7, a particle-wall collision for a rough wall is sketched. In numerical simulations it is usually impossible to include the exact roughness of the wall, because the length scale of wall-roughness is small compared to the grid size. To overcome this limitation, the model of Tsuji et al. [1989] and Sommerfeld [2001] is used. The wall roughness is described by a virtual inclination of the wall at the contact point [Tsuji et al., 1989, Sommerfeld, 2001]. This inclination is represented by an angle β in a 2D case. In a 3D case the inclination can be represented by two angles β_1 and β_2. In Figure 3.8, the virtual wall inclination is sketched. A particle can hit the wall

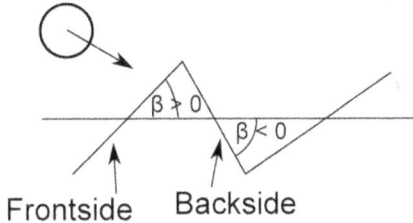

Frontside   Backside

Figure 3.8: Virtual wall inclination angle β. The particle can hit the front or backside of a wall triangle element.

at the front or back side of a wall roughness element. In a first assumption, the inclination angle can be assumed to be Gaussian distributed between $-90° < \beta < 90°$. By considering the problem in more detail, an interesting phenomena is observed. The probability that a particle hits the back is less than the probability to hit the front. By recognizing the impact angle of the particle, a region can be found which can not be hit by the particle. Hence, this region is called shadow region and is sketched in Figure 3.9. Furthermore, if this is considered in the estimation of the inclination angle, the range of possible values for β is decreased. The probability

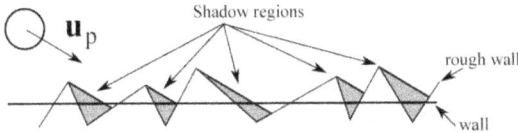

Figure 3.9: Shadow effect in the wall collision model

for frontside collisions increases and the inclination angle distribution changes to a

43

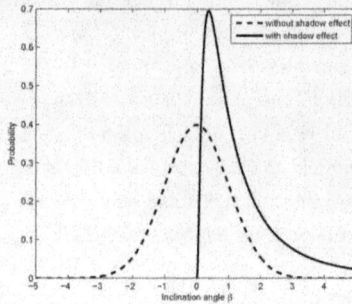

Figure 3.10: If the shadow effect is considered, the probability function for the inclination angle tends to values for front side collision. In this Figure an example is shown, where a back side collision is not possible.

logarithmic behavior (Figure 3.10) [Kahrimanovic, 2009]. At every wall collision, the inclination value is estimated by a random number generator. Realistic values for the inclination angle are small and in the range of a few degrees.

With all the given information, the particle-wall collision can be described as follows. If a particle hits the wall, a wall inclination angle is generated by a random number generator for β_1 and β_2. The wall normal vector is inclined by two rotations with the inclination angles. Hence, the wall collision is computed in a locally inclinated coordination system. For the new local normal vector the particle-wall collision is computed with the given restitution and friction coefficients. Finally, the new particle rotation and velocity is re-transformed to the original vector coordination system given by the global coordinates. The complete algorithm for the particle-wall collision is shown in Appendix B, and can be found in detail in Kahrimanovic [2009] and the literature cited therein.

4

The hybrid model EUgran+Poly

The Eulerian-Eulerian granular phase model and Eulerian-Lagrangian DPM model
are combined into a hybrid model approach EUgran+Poly. Furthermore, models
for a poly-disperse particle distribution and the coupling of these approaches are
discussed.

4.1 Motivation and overview

As already mentioned before, in industrial facilities the amount of particles is very
high. Already, up-to-date high performance computers do not have enough power to
compute the particle track for each particle in an acceptable time. Hence, only the
simulation using Euler-Eulerian granular phase model (Chapter 2) is feasible. However, the disadvantage of the continuous representation is the inefficient modelling of
poly-disperse particle distributions, which can be easily implemented in a Lagrangian

model. Furthermore, the Lagrangian model (Chapter 3) offers a straight-forward implementation to the Magnus force and particle-wall collisions. The Eulerian granular phase approach would require additional transport equations for the particle spin and its fluctuation energy (a possible theory is presented, for example, in Rao and Nott [2008]). This leads to the idea of using a hybrid approach, which combines both methods by computing the forces on a particle in the description that demands less computational effort, either Eulerian-Eulerian or Eulerian-Lagrangian. Following this idea, the Eulerian-Eulerian granular phase interacts with the surrounding fluid and handles the calculation of the Eulerian granular wall interaction, the statistically inter-particle collisions and the impact of the granular pressure (Figure 4.1). The Eulerian-Lagrangian approach is used for the computation of tracer particle trajectories to account for the Magnus force and a detailed resolved particle-wall interaction. The coupling of the two approaches is based on the volume fraction information. In regions with low mass loading, the Eulerian-Lagrangian particles are forced by the gas and are not strongly influenced by the Eulerian granular phase. In regions with high particulate loading, where the information is transported via particle-particle collisions, the coupling forces the tracer particle to follow the solid phase. Hence, the resultant particle movement for the interpretation of the results is given by the Lagrangian tracer trajectories, e.g. velocity of the particle. Furthermore, the volume fraction of the solid phase has to be taken from the Eulerian description.

Pirker et al. [2010] introduced such a hybrid simulation model for CFD simulations of particle-laden flows to fulfil two aims. First, to obtain a model which can accurately compute dense and dilute granular flows within one geometry. Secondly, to improve computational efficiency of simulations for dilute and dense pneumatic conveying systems in industrial applications. Pirker et al. [2010] showed that such a hybrid approach provides better, and more realistic, simulation results when compared to state-of-the-art models. However, this model was developed for mono-dispersed particle-laden flows only and was validated in only one specific situation. For the simulation of the preheating tower, the mono-dispersed model has to be extended to account for poly-disperse particle flow and strand formation.

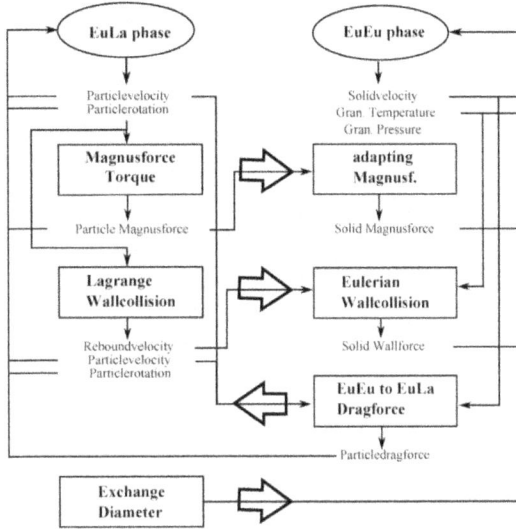

Figure 4.1: Overview of EUgran+Poly model. On the right side is the Eulerian granular phase model with the considered important physical effects depicted. On the left side is the Eulerian-Lagrangian discrete phase model.

4.2 Coupling and exchange forces

The most important part for the hybrid model is the physically correct coupling of the Eulerian and Lagrangian phases. The impact of the forces must be transferred from the Eulerian formulation to the Lagrangian and vice versa. A compatible conversion rule to transfer a force from Lagrangian to Eulerian form can be found by comparing equation (3.2) and equation (2.6) and reads as

$$\mathbf{f}_s = \alpha_s \rho_s \langle \mathbf{f}_p \rangle. \tag{4.1}$$

where subscript s indicates the Eulerian granular phase and subscript p the Lagrangian discrete particle properties. $\langle \mathbf{f}_p \rangle$ denotes a local average of particles property $\mathbf{f}_p$

$$\langle \mathbf{f}_p \left(\langle \mathbf{x}_0 \rangle \right) \rangle = \frac{\sum_{p'} \mathbf{f}_{p'} G \left(\mathbf{x}_{p'} - \mathbf{x}_0 \right)}{\sum_{p'} G \left(\mathbf{x}_{p'} - \mathbf{x}_0 \right)} \tag{4.2}$$

where $G(\mathbf{x}_{p'} - \mathbf{x}_0)$ denotes a weighting function. For a cubic cell, $G = 1$ for $|x_{p'} - x_0| < \frac{\triangle}{2}$ and zero everywhere else. $\triangle$ denotes the grid space. However, in the case of Eulerian-Lagrangian tracer particles, some grid cells may not be hit by a particle trajectory and thus, $\langle \mathbf{f}_p((\mathbf{x}_0)) \rangle = 0$. Hence, for a smooth exchange field, it is important to compute a huge number of Eulerian-Lagrangian trajectories to obtain an area wide exchange field. To reduce the number of Lagrangian tracer trajectories, and for better convergence, smoothing of the exchange field is recommended [Pirker et al., 2011]. One way this can be obtained is by increasing the "radius" of the filter $G(\mathbf{x}_{p'} - \mathbf{x}_0)$, i.e. average over more then one cell.

Most of the standard industry processes involve poly-dispersed particle-laden flows. However, in the Eulerian-Eulerian framework, accounting for poly-dispersity requires additional effort. For each particle diameter class a set of multi-phase Navier-Stokes equations must be solved. Furthermore, a coupling between the different granular phases has to be included. In contrast, it is straight forward to compute Lagrangian trajectories with different particle diameters. These trajectories provide a spatial diameter distribution, which reveals the particle diameter dispersion in the simulation region. In the hybrid approach there is the possibility to compute a particle diameter based on the diameter distribution given from the tracer trajectories. Hence, for each position, a diameter can be found for the Eulerian-Eulerian granular phase by

$$d_s = \langle d_p \rangle = \frac{\sum\limits_{p'} d_{p'} G(\mathbf{x}_{p'} - \mathbf{x}_0)}{\sum\limits_{p'} G(\mathbf{x}_{p'} - \mathbf{x}_0)}. \tag{4.3}$$

However, many forces in the Eulerian granular phase depend non-linearly on the diameter. These forces will have physically wrong values if the arithmetic average of the diameter is used for their computation. Hence, a different coupling between the poly-dispersed tracer particles and the Eulerian-Eulerian phase has to be used. This coupling is based on the average diameter, or the particle diameter distribution, at the specific point in the geometry. Terms in the Eulerian granular model which depend linearly on d_s use the average arithmetic of the diameter (equation 4.3), while all other terms ψ are averaged in each cell, by

$$\langle \psi_p \rangle = \frac{\sum\limits_{p'} \psi_{p'}(d_p) G(\mathbf{x}_{p'} - \mathbf{x}_0)}{\sum\limits_{p'} G(\mathbf{x}_{p'} - \mathbf{x}_0)}, \tag{4.4}$$

where ψ_c denotes the transferred force.

4.3 Coupling forces on the Eulerian granular phase

The Lagrangian phase in the hybrid model is used to get additional data for the Eulerian granular phase. It can be regarded as a detailed look inside the granular phase. This is sketched in Figure 4.2. The Lagrangian part is used to compute the particle rotation and a detailed particle-wall handling. It returns a force to the Eulerian granular phase which represents the average impact of these forces.

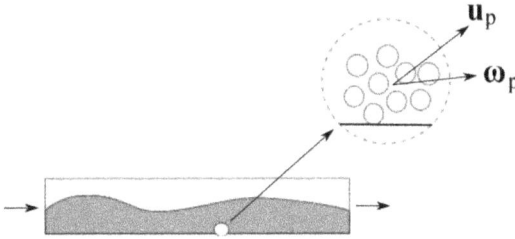

Figure 4.2: The Lagrangian tracer particle cloud gives a detailed look into the granular phase.

4.3.1 Magnus force

The Magnus force and torque were discussed in Section 3.2.2. This force can be easily calculated for Lagrangian trajectories and is transferred by equation(4.1).

4.3.2 Particle-wall interaction

As mentioned in section 3.5.2 the particle-wall collision model of by Kahrimanovic [2009] is being used. It describes the complex wall treatment for particles colliding with the wall. For better agreement of the Lagrangian and Eulerian models, the impact of the Langrangian boundary conditions is transferred to the Eulerian framework with a coupling equation, given by

$$\mathbf{f}_{s,wall} = \alpha_s \frac{\rho_s u_{s,nW}}{\triangle} \left(\langle \mathbf{u}_{p,rebound} \rangle - \langle \mathbf{u}_s + \mathbf{u}_{s,rebound}(\Theta_s) \rangle \right) \tag{4.5}$$

where $\triangle$ denotes the height of the cell. The fluctuating velocity in normal direction to the wall is given by

$$u_{s,nW} = \left| \sqrt{\Theta_s} \mathbf{e}_n \right|. \tag{4.6}$$

The rebound velocity for an average particle in the granular phase is computed by

$$\mathbf{u}_{s,rebound}(\Theta_s) = \sqrt{\Theta_s} \mathbf{e}_{s,rebound} \tag{4.7}$$

where $\mathbf{e}_{s,rebound}$ denotes the normal vector in rebound direction with length of the particle-wall restitution coefficient. The resulting force, $\mathbf{f}_{s,wall}$, is small if the Lagrangian (e. g. Kahrimanovic [2009]) and Eulerian (e. g. Schneiderbauer et al. [2012b]) wall models use the same model for the virtual wall inclination angle. Hence, this force is like a small deflection of the average wall impact force, caused by the velocity fluctuations induced by the granular temperature, Θ_s, and the impact of particle rotation.

4.3.3 Modified drag law for poly-dispersity

In particle conveying and separation in cyclones it can be observed that the solid material tends to form strands. The following main effects can be observed: (i) big particles influence smaller ones to follow them; (ii) the drag between fluid and particle phase depends on the strand diameter and not directly on the particle diameters. The transport of the particles depends on particle-particle collisions and the lee effects of big particles. Only particles at outer positions in the strand are interacting with the surrounding gas. The impact of the gas flow is transported to the inner region of the strand via particle-particle collisions. Due to higher volume fractions inside the particle strand $\mathbf{u}_g$ is nearly the same as the particle velocity $\mathbf{u}_p$ in this regime. Thus a heterogeneous particle distribution is assumed inside the cell, e.g. a particle strand formation of intermediate to dense loading in the cell ($\alpha_s > 10^{-4}$). The drag between fluid and particle can be described for the two regions. Firstly, the drag force, $\mathbf{f}_{p,drag,inside}$, inside the strand is very small and can

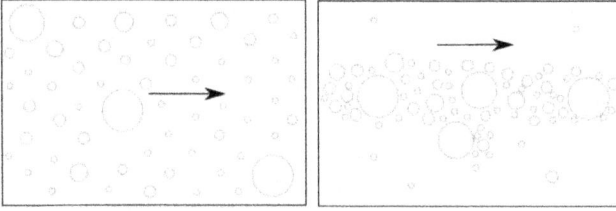

Figure 4.3: Homogeneous and heterogeneous particle distribution in a CFD grid cell.

be neglected. Secondly, the drag depends on the strand diameter, $f_{p,drag,strand}$. The diameter of the strand can be modeled by

$$d_{strand} = K_{strand}d_{p,max}. \tag{4.8}$$

where K_{strand} denotes a constant strand factor which describes the ratio of the strand diameter and the biggest modeled particle diameter. If it is in the range of $K_{strand} = 1 \ldots 1.5$, the influence of the inner particles can be nearly neglected. In this survey we choose $K_{strand} = 1$. CFD simulation results confirm that the particle strands are dominated by the particle strand diameter, which follows the described behaviour. To reduce the impact of the inside smaller particles and knowing that the cross

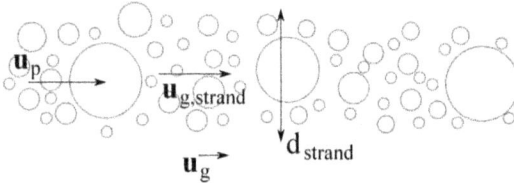

Figure 4.4: Particle strand with different diameters and slipstream effect on gas velocity inside the strand. u_g denotes the computed gas velocity in the grid cell.

area between fluid flow and granular flow is used for modelling the drag force a poly-dispersity drag coefficient for particle diameter class p with

$$K_{poly,p} = \frac{d_p^2}{d_{strand}^2} \tag{4.9}$$

51

is introduced. With this factor, the drag law based on Lagrangian observations is transformed into the Eulerian granular phase, for $10^{-4} < \alpha_s < 0.3$, to

$$\mathbf{f}_{s,drag} = \frac{\sum\limits_{p'} \beta_{poly,p'} \left(\mathbf{u}_g - \mathbf{u}_s\right) G\left(\mathbf{x}_{p'} - \mathbf{x}_0\right)}{\sum\limits_{p'} G\left(\mathbf{x}_{p'} - \mathbf{x}_0\right)} \tag{4.10}$$

with

$$\beta_{poly,p'} = \beta_{p'} K_{poly,p'} \tag{4.11}$$

For a mono-dispersed granular phase, the correction factor, β_{poly}, approaches unity, and the drag force of Wen and Yu [1966] is recovered. For values of α_s outside the given range a standard packing is assumed so that the standard drag law given by Gidaspow et al. [1992] can be used. However, others [eg, van der Hoef et al., 2008] suggest that the correction factor has the form of a third order polynomial. Nevertheless, our simulations show that the dominant term is the quadratic term since the drag force depends on the area normal to the flow direction, which justifies equation (4.9). For the cases of the preheating tower and the inside of cyclones, these effects are mainly responsible for the particle movement behavior and separation efficiency, which can be seen later in the simulation results (Chapter 7.1.2).

4.4 Coupling forces on the Lagrangian tracer particles

The main tasks of the Eulerian granular phase are the modelling of the solid phase and the force exchange with the surrounding gas. Additionally, it provides information for the computation of source terms influencing the Lagrangian tracer particles. More specifically, these forces are a particle-solid drag force and a particle granular pressure force. In Figure 4.1, these two forces are combined in the module "EuEu to EuLa drag force".

4.4.1 Collisional particle-solid force

As already mentioned, in industry most of the commonly used materials are poly-dispersed. The collisional particle-solid force describes the effect of particle move-ment due to the momentum exchanging collisions between particles of different diam-eters. Obviously, it depends on the volume fraction. Furthermore, this force is at the heart of the coupling between solid phase and Lagrangian tracer particles. In very dilute regions, the Lagrangian tracer particles are nearly free of the effect of the solid phase and follow the gas flow. In dense regimes, where particle collisions dominate, the Lagrangian tracers are forced to follow the solid phase to include the influence of collisions to the Lagrangian description. Thus, the collisional particle-solid force must decrease with decreasing volume fraction, α_s, and vice versa. A model for this momentum exchange via collisions, for the case of poly-dispered modelling in the Eulerian description, is already present in FLUENT [2006]. This momentum exchange between particles with different diameters and velocities is modeled with a solid-particle exchange coefficient for poly-dispersed granular. It reads [FLUENT, 2006]

$$\beta_{p,coll} = \frac{3\left(1 + e_{s,p}\right)\alpha_s\left(d_p + d_s\right)^2}{4\left(d_p^3 + d_s^3\right)}g_0\left|\mathbf{u}_s - \mathbf{u}_p\right|. \tag{4.12}$$

where d_p denotes the actual particle trajectory diameter and d_s the solid diameter in the actual grid cell. Due to the higher number of collisions inside the strand, a modification is required. This modification is for the case of strand formation due to the change in the particle-particle collision drag. Therefore, a simple manipulation of the equation is suggested for the region between $10^{-1} < \alpha_s < 0.3$ given by

$$\beta_{p,coll} = \frac{3\left(1 + e_{s,p}\right)\sqrt{0.2\alpha_s}\left(d_p + d_s\right)^2}{4\left(d_p^3 + d_s^3\right)}g_0\left|\mathbf{u}_s - \mathbf{u}_p\right|. \tag{4.13}$$

to account for the increasing number of collisions inside the particle strand.

4.4.2 Granular pressure force

The granular pressure force is based on the observation that particles avoid regions that are already densely packed. Particles in dense regions tend to move to dilute regions due to particle-particle collisions. Additionally, it can be observed that particles which hit a particle-gas surface are reflected [Pirker et al., 2010]. The

53

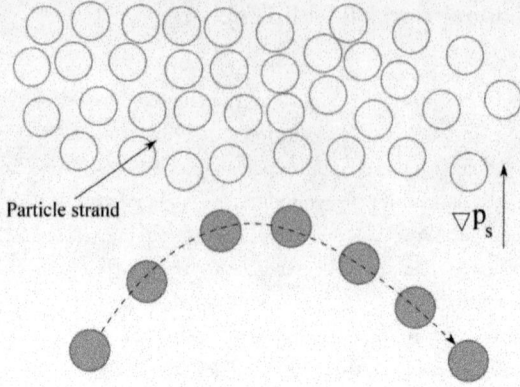

Figure 4.5: Illustration of granular pressure.

granular pressure force can be calculated by the gradient of the volume fraction field, α_s, or the gradient of the granular pressure field, p_s. Hence, the force based on granular pressure is given by

$$\mathbf{f}_{p,press} = -\nabla p_s \frac{1}{\alpha_s \rho_s} \qquad (4.14)$$

and tracer particles that move against an increasing granular pressure field are forced to move back. In Fig. 4.5 the force is illustrated and a particle's deflection is sketched.

4.4.3 Collisional torque

The influence of particle-particle collisions on the particle rotation can not be neglected and should be considered. Therefore, a collisional torque in the rotational momentum balance of the Lagrangian system is added. The additional torque term in equation (3.3) can be written as

$$\mathbf{t}_{p,add} = -C_R \dot{n}_{s,p} \boldsymbol{\omega}_p, \qquad (4.15)$$

with the estimated particle collision frequency

$$\dot{n}_{s,p} = 4 n_s n_p d_p d_s g_0 \sqrt{\pi \Theta_s}, \qquad (4.16)$$

54

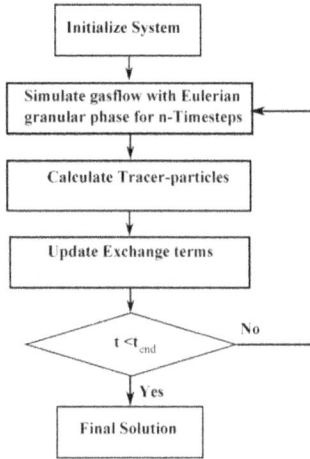

Figure 4.6: The flowchart for the simulation with the hybrid model. An Eulerian granular phase simulation is computed for N time steps. After that, the tracer particles are computed with additional information gained by the values of the solid phase. After their computation, the source force fields from the Langrangian tracer are computed and smoothed. After that the Eulerian granular simulation and the whole process is redone.

where n_s is the number of granular phase particles and n_p is the number of particles in the actual set of tracked DPM particles. C_R is a constant which is around one and allows a fine tuning of this simple collisional torque model [Pirker et al.. 2010].

4.5 Simulation sequence and implementation

After discussion of the idea and the exchange forces a flow chart of the model implementation is given (Figure 4.6). The hybrid model consists of a standard Eulerian granular phase simulation with additional force terms. The computation of the Eulerian granular phase is interrupted after a fixed number of time steps (Figure 4.6). This number has to be chosen with respect to the simulated problem and should be in a range such that the Eulerian granular phase model can react to the changing forces induced by the Lagrangian tracers and vice versa. Good agreement was found in the region of 100 to 500 timesteps. Then, the velocities of the fixed solid and gas fields are used to calculate the Lagrangian tracer particles trajectories. During the calculation, the impact of the Eulerian granular phase is considered by computing the source terms in each cell that is hit by a tracer particles. For each particle diameter class, a set of tracer trajectories is computed. After their computation, the source forces for the Eulerian granular phase are computed in each grid cell and smoothed over the computational domain. After that, the computation of the Eulerian granular phase continues and the whole process is restarted. However, between the computational step before and after the computation of the Lagrangian tracer particles, the exchange fields (forces and diameter) change their values promptly. These changes are a problem for the solver stability and therefore the convergence quality. Hence, to get a smooth change of the exchange fields, it is suggested to use a linear interpolation over K_{smooth} timesteps. Simulations showed, that stability can be guaranteed if the change of the diameter between two timesteps is less than one percent and therefore $K_{smooth} > 100$.

The hybrid model is implemented in a modular way with User Defined Functions (UDF) in FLUENT. Hence, it is possible to change every model in a fast and simple way. The UDF is written in C-code and is ready for parallel computation. The UDF must be compiled on the same machine where it is used in FLUENT. The UDF structure is presented in Appendix C.

To have no time for philosophy is
to be a true philosopher.

Blaise Pascal (1623-1662)

5

Agglomeration

Agglomeration describes the process of aggregation of two or more colliding particles (Figure 5.1). Agglomeration changes particle properties like size, surface and mass.

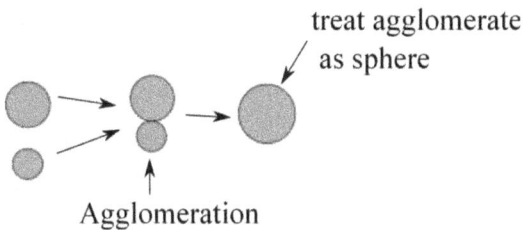

Figure 5.1: Simple sketch of the assumed agglomeration process

Therefore, the physical behavior of particle aggregates in a flow change. Thus, the effect of agglomeration is important for particle transport and particle separation processes. In particle-laden flows many agglomeration mechanisms are observed. These are: agglomeration based on Brownian motion, fluid kinematics, turbulence, triboelectric forces, acoustics and other forces. Agglomeration based on **Brownian motion** depends on the temperature. With increasing temperature, the fluctuation of very small particles increases [Sitarski and Seinfeld, 1977]. In the free molecular or transition regime (Knudsen number Kn = 0.8 − 1.6), the inter-particle van der Waals attraction is present [Kerminen, 1994]. Depending on the Brownian motion and the van der Waals attraction, there is a chance that particle collision and sticking can occur. Agglomeration based on Brownian motion is often called perikinetic agglomeration. Agglomeration based on **fluid kinematics** depends on the shear gradient in the turbulent gas. Furthermore, induced by this slip velocity, particles move with different speeds. Based on this velocity difference between particles, a collision probability and agglomeration rate can be computed. This kinematic agglomeration effect is important for building aggregates with diameters bigger than $1\,\mu$m [Anh, 2004]. The **turbulence** in the gas induces a fluctuation force onto the particles, which cause the particle motion to fluctuate. This increases the collision probability of particle aggregates and therefore the probability for an agglomeration event. Based on particle-particle, particle-wall and gas-wall shear forces, a triboelectric force can be present in the system. This force has a direct impact on the adhesive forces between particles. Additionally, in some gas flows, acoustic waves with high frequency are present. These induce a harmonic oscillating motion on the particles. Hence, the collision frequency between the particles is increased. All forces which do not depend directly on effects happening in the flow process are referred to external forces. This can be an electric field, a magnetic field, a field of gravity or a combination of them. Agglomeration has an important effect on the separation efficiency of cyclones, because the separation efficiency increases with increasing average particle diameter. For the case of the preheating tower, the most important factors affecting agglomeration are fluid kinematics, Brownian motion and turbulence. From measurements, it could be observed that a triboelectric force may also be present. However, from further validaton, it was found the effect may depend on the materials used in the lab-scale experiment, and may not be present in the full-scale plant. Hence, the triboelectric, acoustic and external forces are not considered in this study. For a more detailed discussion of this topic, Meyer and Deglon [2011] and Park et al. [2002] present a very brief introduction and review on agglomeration. For actual research standards, the literature of Zaichik et al. [2006], Zaichik and Alipchenkov [2006], and literature cited therein, are suggested.

5.1 Simple models

As this study deals mostly with industrial problems, some fast and cheap models will initially be presented. If these give accurate enough results for a given problem, then more complex models with higher computational expense can be neglected.

We assume that the conveyed or separated material consists of particles with diameter d_i. The number of different diameters is given by N. Hence, the material consists of N diameter classes. The amount of each diameter, d_i, in the material is given by h_i and

$$\sum_{i=1}^{N} h_i = 1. \tag{5.1}$$

5.1.1 Agglomerated filling

The simplest, and most computational efficient, approach is to feed the system with already agglomerated material. This is only possible if, the agglomeration takes place directly at the point of material feeding in the plant. This would be the same as if already-agglomerated material is inserted. Obviously, for this model, the particle size distribution in the agglomerated material has to be known.

5.1.2 Linear agglomeration

Secondly, if the change of the particle diameter distribution from the entry to the outlet of the device is known, an interpolation model can be used. Therefore, a linear or polynomial interpolated transfer function is used. This transfer function computes the agglomeration of the material over the time. Linearly interpolated agglomeration for the i-th diameter class can be given with,

$$h_i(t) = h_i(t_0) + \frac{h_{i,\text{agg}} - h_i(t_0)}{t_{\text{agg}} - t_0}(t - t_0) \qquad t_0 \leq t \leq t_{\text{agg}} + t_0. \tag{5.2}$$

where t_0 denotes the time at the beginning of the agglomeration process, t_{agg} the duration of the agglomeration process, $h_i(t_0)$ denotes the amount of the i-th particle class at the inlet and $h_{i,\text{agg}}$ the agglomerated amount of the i-th particle class.

Therefore, for this model, the following parameters must be known: t_{agg}, $h_i(t_0)$ and $h_{i,\text{agg}}$. These values have to be found by experiments.

5.2 Particle population balance equation

In the early 20-th century Smoluchowski [1917] presented a population balance equation (PBE) for modelling agglomeration. This PBE describes the distribution of particle diameter classes. The given equation system can be solved with special requirements for the particle classes.

5.2.1 Assumptions

The model of Smoluchowski [1917], which consists of the PBE and models for the collision rates, is based on the assumptions discussed below (e. g. Meyer and Deglon [2011]). The particles, as well as the aggregates of particles, are spherical. Initially, only agglomeration is considered, while no aggregate breakup is accounted for. Furthermore, only binary collisions between the particles are assumed, which is often not the case in frictional regimes. For the case of the preheating tower, where the collisional regime dominates, the influence of the frictional regime can be neglected. Furthermore, in this book, the impact of friction on the agglomeration is not considered. However, for the i-th particle class, the population balance is given by,

$$\frac{\partial n_i(t)}{\partial t} = \frac{1}{2} \int_0^i K_{j,i-j}(t) n_j(t) n_{i-j}(t) \mathrm{d}j - n_i(t) \int_0^\infty K_{i,j}(t) n_j(t) \mathrm{d}j - n_i(t) R_i(t) + S_i(t),$$

(5.3)

where index j denotes the agglomeration partner, and $n_i(t)$ denotes the number of particles per cubic meter with diameter d_i at time t. $K_{i,j}$ denotes the exchange rate of particle classes i and j, $R_i(t)$ is the deposition rate and $S_i(t)$ the production rate of particles in class i. In this work, no modelling of deposition and production is needed, so these two rates are neglected. Smoluchowski [1917] assumed that an aggregate of particles i consists of two particles j and $i - j$ ($i > j$). Furthermore, particle i is treated as a spherical particle with the mass of the two agglomerated particles j and $i - j$. Hence, if particle i agglomerates again with another particle, it is treated as a particle and not as an aggregate of particles. After discretization for N particles classes, the equation changes to

$$\frac{\partial n_i(t)}{\partial t} = \frac{1}{2} \sum_{j=i}^{i-1} n_{i-j}(t) K_{i-j,j}(t) n_j(t) - n_i \sum_{j=i}^{N} K_{i,j}(t) n_j(t). \tag{5.4}$$

Furthermore, it must be guaranteed that $V_{p,i-j} + V_{p,j} = V_{p,i}$ (Figure 5.2). This must be accounted for by choosing the particle diameter distribution. Otherwise, there would be a problem where agglomeration creates or destroys mass. Therefore, because no volume should be created by agglomeration, the particle diameter distribution can be calculated by

$$V_{p,i-j} + V_{p,j} = V_{p,\text{agg}} \tag{5.5}$$

$$\frac{d_{i-j}^3 \pi}{6} + \frac{d_j^3 \pi}{6} = \frac{d_{\text{agg}} \pi}{6} \tag{5.6}$$

$$d_{i-j}^3 + d_j^3 = d_{\text{agg}}^3 \tag{5.7}$$

$$\left(d_{i-j}^3 + d_j^3 \right)^{\frac{1}{3}} = d_{\text{agg}}. \tag{5.8}$$

Introducing the requirement that the minimum modeled diameter is given by $d_j = j^{\frac{1}{3}} d_{\min}$ the equation changes to

$$\left((i - j) d_{\min}^3 + j d_{\min}^3 \right)^{\frac{1}{3}} = d_{\text{agg}} \tag{5.9}$$

$$i^{\frac{1}{3}} d_{\min} = d_{\text{agg}}. \tag{5.10}$$

Hence, the particle diameter distribution is found from

$$d_i = i^{\frac{1}{3}} d_{\min}. \tag{5.11}$$

Using this rule would cause another problem. Between diameter $d_{\min}$ and $10 d_{\min}$ 1000 different diameters must be used to fill the range. Often a factor of more than 10000 is present between $d_{\min}$ and $d_{\max}$ in industrial applications. Obviously, due to excessive computational expense, this diameter distribution is not practical. Hence, another approach to the solution of the balance equation has to be found, e. g. the

61

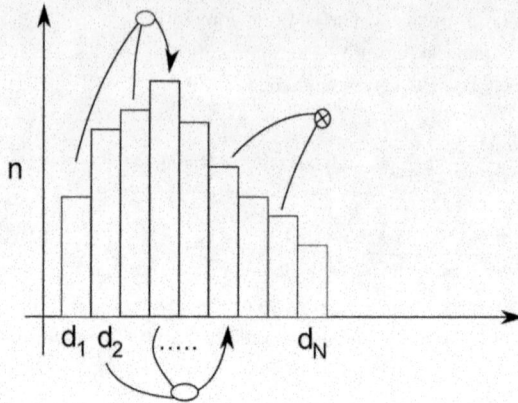

Figure 5.2: Discretized particle diameter distribution. The arrows show in which class two agglomerating particles will create a new particle. $\otimes$ indicates that aggregates which are bigger than the threshold diameter are not considered.

method by Hounslow et al. [1988]. The growth rate of a particle diameter class is described by [Kumar and Ramkrishna, 1997]

$$\frac{\partial n_i(t)}{\partial t}\Big|_{\text{growth}} = G_i \left(a n_{i-1} + b n_i + c n_{i+1} \right) \tag{5.12}$$

where G_i denotes the growth rate for particles of size V_i. a, b and c are coefficients which have to be calculated in such a way to enforce that the given equation yields the correct expression for three moments [Kumar and Ramkrishna, 1997]. Hounslow et al. [1988] mentioned that negative populations would be possible with his model. This must be avoided by explicitly setting them to zero. Kumar and Ramkrishna [1997] revised the study of [Hounslow et al., 1988]. In addition, when considering Figure 5.2 the maximum diameter $d_N = d_{\text{max}}$ should be in a range such that stable aggregates of this size can not be created. This would avoid the problem that the agglomeration model must include a threshold for maximum creatable aggregates. In real applications this happens because of de-agglomeration. In this book, we only use a threshold diameter, because de-agglomeration is not considered.

For the calculation of the agglomeration, the agglomeration rate coefficients, $K_{i,j}$, are needed. These coefficients describe the kinetics of the agglomeration process. The agglomeration rate between two particles from class i and j is given by

$$K_{i,j} = N_{i,j}\eta_{i,j}H_{i,j} \tag{5.13}$$

where $N_{i,j}$ denotes the collision rate without accounting for the surrounding fluid, $\eta_{i,j}$ is the correction of the collision probability due to fluid effects and $H_{i,j}$ denotes the sticking probability.

5.2.2 Collision rates

The collision rate describes the rate of colliding particle volumes. Obviously, a collision between particles can only occur if the particles move with different velocities. This depends on the following effects [Park et al., 2002]: the velocity gradient in the fluid phase, Brownian motion, fluid turbulence, kinematic particle movement, electrostatic agglomeration and also acoustic agglomeration. For each of these effects, a collision rate can be computed. The sum of all these effects reveals the total collision rate. As mentioned before, only agglomeration based on Brownian motion, turbulence and kinematic effects is considered in this study.

5.2.2.1 Kinematic collision rate

For two particle classes, i, and j, moving with different velocities, a collision frequency can be computed. It is given by [Kruis and Kuster, 1997]

$$N_{i,j} = \sqrt{\frac{8\pi}{3}} \left(r_{p,i} + r_{p,j}\right)^2 \sqrt{\left(u_{p,i} - u_{p,j}\right)^2} \tag{5.14}$$

where $r_{p,i}$ denotes the radius of particle i. For particles with identical velocities, the collision rate is zero. Nevertheless, the collision rate is important in regions where particles with different diameter have considerably different velocities. For each particle class in a grid cell, a velocity is given. Hence, particles with identical diameter move with the same velocity and no agglomeration by kinematics is present.

5.2.2.2 Brownian collision rate

For very small particles (μm), the Brownian motion of the particles becomes non-negligible. Brownian motion is treated as a velocity fluctuation of the particles, from which a collision rate can be calculated. For the description of the flow regime [Sitarski and Seinfeld, 1977], the Knudsen number for particles, given by

$$\mathrm{Kn} = \frac{2\lambda}{d_p} \tag{5.15}$$

is used. The Knudsen number describes the ratio of molecular mean free path and particle diameter. It is a dimensionless number for length characterization. λ denotes the mean free path and is given by

$$\lambda = \frac{k_B T}{\sqrt{2}\pi\sigma^2 p}, \tag{5.16}$$

where T denotes the gas temperature, σ the gas molecule hard shell diameter, p the total pressure and k_B the Boltzmann constant. In particle-laden flows with the Knudsen number $\ll 1$, the following equation for the collision rate is given [Smoluchowski, 1917, Meyer and Deglon, 2011]

$$N_{i,j,\mathrm{brownian}} = \frac{2k_B T}{3\mu}\left(\frac{1}{r_{p,i}} + \frac{1}{r_{p,j}}\right)(r_{p,i} + r_{p,j}). \tag{5.17}$$

This is also the case for flow situations, which are considered in this book. For higher Knudsen numbers, other models have to be used [Anh, 2004].

5.2.2.3 Turbulent collision rate

The turbulence in the gas induces a fluctuation force on the particles. The resulting particle fluctuations determine the probability of particle aggregates. The interaction of the particle and turbulence can be characterized by the Stokes number (Figure 5.3). The particle Stokes number describes the ratio of particle relaxation time and the characteristic timescale for gas turbulence, which is also referred to as the Kolmogorov timescale T_t. For the k-th particle, the relaxation time is given by

$$\tau_k = \frac{(2\rho_p + \rho_f)\,r_k^2}{9\mu}, \tag{5.18}$$

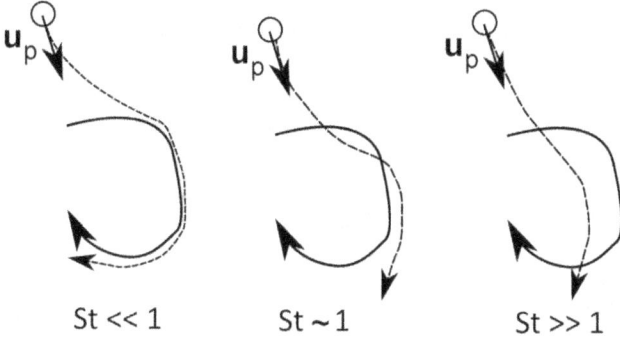

Figure 5.3: Influence of turbulence to the particle movement by Stokes number.

where r_k denotes the particle radius and μ is the dynamic viscosity of the fluid. Commonly, $\rho_p \gg \rho_f$ and ρ_f can be neglected. The equation changes to

$$\tau_k = \frac{\rho_p d_k^2}{18\mu}. \tag{5.19}$$

The Stokes number is given by

$$\mathrm{St}_{p,k} = \frac{\tau_k}{T_t}. \tag{5.20}$$

For different particle Stokes numbers, different particle-laden flow are observed. Therefore, different models for the particle collision rate were developed. If the particles are smaller than the turbulent eddies, or a high Kolmogorov timescale is present, the dynamic movement of the particles is dominated by the turbulent dissipation rate and the kinematic viscosity of the gas [Anh, 2004]. This is observed for Stokes number $\mathrm{St}_i \to 0$. A solution for the collision rate for very small Stokes numbers was given by Saffman and Turner [1956] with

$$N_{i,j} = \left(\frac{8\pi}{15}\right)^{\frac{1}{2}} (r_i + r_j)^2 \left[3\left(1 - \frac{\rho_f}{\rho_p}\right)^2 \times (\tau_i - \tau_j)^2 \overline{\left(\frac{\mathrm{D}v_f}{\mathrm{D}t}\right)^2} + \frac{1}{5}\left(\frac{\epsilon}{\nu}\right)^2\right]^{\frac{1}{2}}. \tag{5.21}$$

In Saffman's equation, ϵ, is the dissipation rate of the turbulent energy, ν is the kinematic viscosity of the fluid and τ_k is the particle relaxation time. Furthermore, the average acceleration of eddies in the dissipation range is needed and it is given by [Hinze, 1975]

$$\overline{\left(\frac{Dv_f}{Dt}\right)^2} = 1.16\epsilon^{\frac{3}{2}}\nu^{-\frac{1}{2}}. \tag{5.22}$$

If the particles are big compared to the eddies, or a small Kolmogorov timescale is present, the particles often travel from one eddy to another. There the Stokes number increases to a maximum of $St_i \to \infty$. Furthermore, in these regions the particles are imparted a small, but high frequency, fluctuation. To account for this situation, Abrahamson [1975] derived a collision rate function for high Stokes numbers from kinetic gas theory and particle velocity distribution. The threshold for collision rates at $St_i \gg 1$ is given by,

$$N_{i,j} = 2^{\frac{3}{2}}\pi^{\frac{1}{2}}\left(r_{p,i} + r_{p,j}\right)^2 \sqrt{\nu_{p,i}^2 + \nu_{p,j}^2} \tag{5.23}$$

where $\nu_{p,i}^2$ denotes the mean square velocity for particle i. An estimation of the mean square particle velocity was given by Abrahamson [1975] with

$$\overline{\nu_{p,k}^2} = \frac{\overline{\nu_f^2}}{1 + 1.5\tau_{p,k}\epsilon\overline{\nu_f^2}}. \tag{5.24}$$

After the derivation of solutions for high and low Stokes numbers, a description for the collision rate in the intermediate range has to be found. Since the particle movement is dependent on the eddies induced by viscous and kinematic forces, these effects must also be considered. The first model for the whole range of Stokes numbers was presented by Williams and Crane [1983]. It is given by

$$N_{i,j} = \sqrt{\frac{8\pi}{3}}\left(r_{p,i} + r_{p,j}\right)^2 \sqrt{\overline{\nu_f^2}\frac{3}{(St_i + St_j)^2 - 4St_iSt_j\sqrt{\frac{1+St_i+St_j}{(1+St_i)(1+St_j)}}}{(St_i + St_j)(1 + St_i)(1 + St_j)}}. \tag{5.25}$$

This model excludes the effect of the shear mechanism, which means the results do not match the correct results of Saffman and Turner [1956] for the accelerative mechanism in the small particle limit [Park et al., 2002]. Therefore, Kruis and Kuster [1997] modified the model to deal with the shear mechanism. The model of Kruis and Kuster [1997] can also be used for the complete range of particle Stokes numbers [Park et al., 2002], and is therefore used in this survey. The model by

Kruis and Kuster [1997] consists of the velocity of acceleration, w_a^2, and the velocity of shear gradient, w_s^2. The complete collision rate is described by

$$N_{i,j} = \left(\frac{8\pi}{3}\right)^{\frac{1}{2}} (r_i + r_j)^2 \left(w_a^2 + w_s^2\right)^{\frac{1}{2}}. \tag{5.26}$$

Park et al. [2002] and Meyer and Deglon [2011] mention that the acceleration part of the relative particle velocity, w, is represented by,

$$w_a^2 = 3(1-\delta)^2 \nu_f^2 \frac{\gamma}{\gamma-1} \frac{(\text{St}_i + \text{St}_j)^2 - 4\text{St}_i\text{St}_j \sqrt{\frac{1+\text{St}_i+\text{St}_j}{(1+\text{St}_i)(1+\text{St}_j)}}}{\text{St}_i + \text{St}_j}$$
$$\times \left\{ \frac{1}{(1+\text{St}_i)(1+\text{St}_j)} - \frac{1}{(1+\gamma\text{St}_i)(1+\gamma\text{St}_j)} \right\} \tag{5.27}$$

where γ is a spectrum constant and can be calculated with

$$\gamma = 0.183 \frac{\nu_f^2}{\sqrt{\epsilon\nu}}. \tag{5.28}$$

γ typically has a value between 10 and 100 [Park et al., 2002]. The turbulent energy, k and dissipation rate ,ϵ, are usually known from CFD simulations, hence γ can be directly computed. The shear part is given by,

$$w_s^2 = 0.238\delta\nu_f^2 \left(\frac{\nu_i^2}{\nu_f^2} \frac{\text{St}_i}{C_{c,i}} + \frac{\nu_j^2}{\nu_f^2} \frac{\text{St}_j}{C_{c,j}} + 2\frac{\overline{\nu_i\nu_j}}{\nu_f^2} \sqrt{\frac{\text{St}_i\text{St}_j}{C_{c,i}C_{c,j}}} \right) \tag{5.29}$$

where

$$\delta = \frac{\rho_f}{2\rho_p + \rho_f} \tag{5.30}$$

denotes a mass coefficient and $C_{c,i}$ denotes the Cunningham correction factor of particle i. ν_i, ν_j and ν_f denote the root mean square velocities of particles i and j and the fluid f. The Cunningham correction factor [Cunningham, 1910] or Cunningham slip correction factor is used to account for non-continuum effects when calculating the drag on small particles. The derivation of Stokes Law, which is used to calculate the drag force on small particles, assumes a no-slip condition that is no longer correct at high Knudsen number. The Cunningham slip correction factor allows the prediction of the drag force on a particle moving in a fluid with a Knudsen number between the continuum regime and free-molecular flow. Small Knudsen numbers are expected in the particle transport regime of the preheating tower, and therefore the Cunningham slip correction factor is $C_c = 1$. In equation (5.29), the root mean square particle velocity is given by [Park et al., 2002]

$$\frac{\nu_i^2}{\nu_f^2} = \frac{\gamma}{\gamma - 1} \left[\frac{1 + \delta^2 St_i}{1 + St_i} - \frac{1 - \delta^2 \gamma St_i}{\gamma \left(1 + \gamma St_i\right)} \right] \tag{5.31}$$

for the i-th particle. The correlation of the velocities is given by

$$\frac{\nu_i \nu_j}{\nu_f^2} = \frac{\gamma}{\gamma - 1} \left[\frac{\left(St_i + St_j + 2St_i St_j\right) + \delta \left(St_i^2 + St_j^2 - 2St_i St_j\right)}{\left(St_i + St_j\right)\left(1 + St_i\right)\left(1 + St_j\right)} \right.$$
$$+ \frac{\delta^2 \left(St_i^2 St_j + St_i St_j^2 + 2St_i St_j\right)}{\left(St_i + St_j\right)\left(1 + St_i\right)\left(1 + St_j\right)}$$
$$- \frac{\left(St_i + St_j + 2\gamma St_i St_j\right) + \delta\gamma \left(St_i^2 + St_j^2 - 2St_i St_j\right)}{\gamma \left(St_i + St_j\right)\left(1 + \gamma St_i\right)\left(1 + \gamma St_j\right)}$$
$$\left. - \frac{\delta^2 \gamma^2 \left(St_i^2 St_j + St_i St_j^2 + \frac{2}{\gamma} St_i St_j\right)}{\gamma \left(St_i + St_j\right)\left(1 + \gamma St_i\right)\left(1 + \gamma St_j\right)} \right]. \tag{5.32}$$

All of the presented models are based on the consideration of correlated or independent acceleration or orthokinetics. Newer models, which are not included in this book, are based on preferential particle concentration. This type of modelling was introduced in the work of Reade and Collins [2000]. Zaichik et al. [2006] and Zaichik and Alipchenkov [2006] introduced an equation based on preferential concentration for the particle collision frequency. A detailed review on the history of turbulent collision rates can be found in Meyer and Deglon [2011]. In industrial facilities, in particular the preheating tower, it is common to consider the entire range of particle Stokes numbers. Thus the model of Kruis and Kuster [1997] has been commonly used in the industry over the last few years. Therefore it was decided to use this model instead of switching to the newer approaches.

5.2.2.4 Comparison of collision rates

In Figure 5.4, a comparison of the agglomeration based on Brownian motion and turbulent motion [Kruis and Kuster, 1997] is plotted. The sticking probability, $H_{i,j}$, and the collision probability correction, $\eta_{i,j}$, are set to one. Hence, the agglomeration rate, $K_{i,j}$, is identical to the collision rate, $N_{i,j}$, computed for Brownian and turbulent assumptions. For comparison, a particle with diameter $d_i = 1\,\mu m$ is used, and collides with particles with diameters from 0.01 to $10^4\,\mu m$. Furthermore, the Stokes number for these particles is between $St_j \ll 1$ and $St_j \approx 25500$. For particles

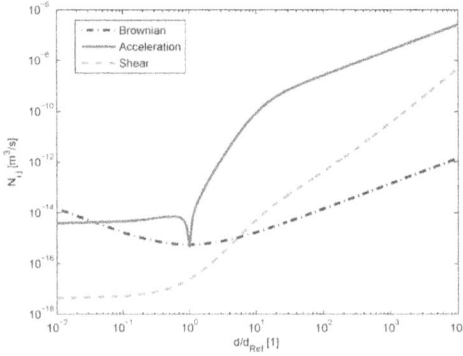

Figure 5.4: Schematic diagram, showing the agglomeration rate of the different models for a $d_{Ref} = 1\,\mu m$ and $St_j = [0.25500]$. The impact of Brownian agglomeration decreases with increasing diameter, while the agglomeration based on acceleration and shear increases.

that are much smaller than $1\,\mu m$, the Brownian agglomeration rate is important. This is because the influence of Brownian motion is many orders of magnitude higher than the collision rate due to shear. For particles with larger diameters, the turbulent agglomeration becomes more important. Therefore, for particle diameter ratios $\frac{d_{p,i}}{d_{p,j}} > 10$ ($d_{p,i} > d_{p,j}$), the Brownian motion can be neglected to save computational time. For particles with the same diameter, the acceleration collision frequency tends to zero. This because particles with the same diameter are predicted to have the same local velocity.

5.2.3 Effective collision rate

The effective collision probability, $\eta_{i,j}$, describes the correction of the collision probability due to the fluid surrounding the particle. It can be observed that particles are influenced by the fluid stream around the particle. Furthermore, the particle can be forced to go with the fluid around the collision partner, without colliding (Figure 5.5). Schuch and Löffler [1975] presented a model to include this effect to the agglomeration rate. It is a model that gives a probability between zero and one that a particle and the collision partner are really colliding. The collision efficiency

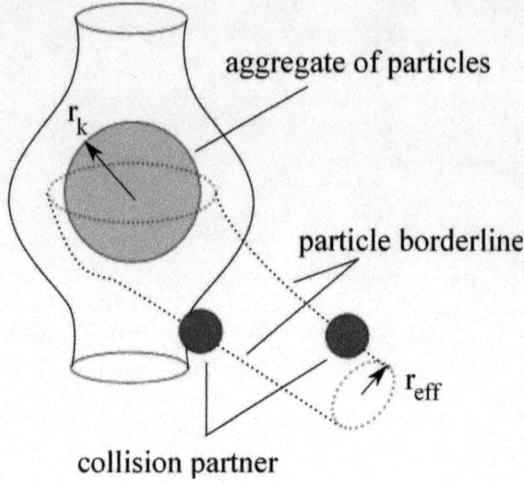

Figure 5.5: Schematic illustration of collision in a turbulent fluid flow. If the collision partner is inside a defined area before collision, the collision happens. Otherwise the particle flows around the collision partner, due to the effect of the surrounding fluid.

is calculated based on the ratio of the collision area to the projected area of the aggregated particles [Anh, 2004]. It is given by

$$\eta_{i,j} = \frac{r_{\text{eff}}^2 \pi}{r_k^2 \pi} \tag{5.33}$$

where $r_{\text{eff}}^2 \pi$ denotes the area of the effective collisions area and $r_k^2 \pi$ the projected area of the aggregated particle. Schuch and Löffler [1975] observed the collision efficiency between $5\,\mu\text{m}$ limestone particles and $100\,\mu\text{m}$ glycerin-water droplets. With numerical computations of two dimensional flow fields, they derived a equation for the estimation of the collision efficiency [Anh, 2004] based on the Stokes number given by

$$\text{St}_{\text{rel}} = \frac{\rho_p |\mathbf{u}_{\text{rel}}| d_p}{18 \mu_f d_k}. \tag{5.34}$$

The collisions efficiency is then

$$\eta_{i,j} = \frac{r_{\text{eff}}^2 \pi}{r_k^2 \pi} = \left(\frac{\text{St}_{\text{rel}}}{\text{St}_{\text{rel}} + \xi} \right)^{\zeta} \tag{5.35}$$

where ξ and ζ represent parameters which must be chosen based on the particle Reynolds number [Sommerfeld, 2000]. For $Re_p < 1$, they are given by

$$\xi = 0.65 \qquad \zeta = 3.7 \qquad\qquad (5.36)$$

and for $Re_p \gg 1$

$$\xi = 0.25 \qquad \zeta = 2.0. \qquad\qquad (5.37)$$

The presented values are a good choice for the simulation of the collision efficiency based on experiments with scrubbers. For the case of the preheating tower, values for ξ and ζ should be gained through experiments. Furthermore, collision efficiency in turbulent flows between particles and droplets, as well as between particles and particles, is still under investigation. Different models were presented by de Almeida [1979] and Pinsky et al. [1999]. A theoretical formulation for agglomeration of droplets in turbulent flow was formulated by Wang et al. [2005]. Pinsky et al. [1999] showed that the effective collision area is not a circular area, it is better described with an ellipsoid. With this approach, he computed a higher collision efficiency in turbulent flow compared to laminar flow around the particles. Considering the industrial background of the simulations, the model presented here is accurate enough.

5.2.4 Sticking probability

The sticking probability, $H_{i,j}$, describes if two colliding particles will stick together. Therefore it can be defined by a simple description

$$\begin{aligned} H_{i,j} &= 0 \qquad \text{adhesion criterion not fulfilled,} \\ H_{i,j} &= 1 \qquad \text{adhesion criterion fulfilled.} \end{aligned} \qquad (5.38)$$

This means that an adhesion criterion must be found that describes the adhesion force between the two particles. If the force, based on the escape speed, is smaller than the adhesion force, the particles will stick together. The different adhesive forces depend on the particle diameter, e.g. the van der Waals force depends on d_p^2, and the Coulomb force on $\frac{1}{d_p^2}$. Hiller and Löffler [1980] suggested the creation of an energy balance. The kinetic energy, E_{before}, of the particles before the particle-

71

particle collision is the sum of energy after collision, E_{after}, and the energy dissipated by mechanical deformation, E_{mech}, and van der Waals force, E_{vW} .

$$E_{before} = E_{after} + E_{vW} + E_{mech} \qquad (5.39)$$

From this balance, a critical velocity can be derived that describes the minimum impact velocity of the particles at which rebound will occur. It was given by Hiller and Löffler [1980] with

$$u_{crit} = \frac{1}{d_p} \frac{\sqrt{1 - e_p^2}}{e_p^2} \frac{A}{\pi z_0^2 \sqrt{6 p_p \rho_p}} \qquad (5.40)$$

where e_p denotes the coefficient of restitution, A is the Hamaker constant and z_0 describes the minimum particle-particle contact distance [Anh, 2004]. Hiller and Löffler [1980] demonstrated that the estimated critical velocity for particles in the range of $1\,\mu\text{m} \leq d_p \leq 10\,\mu\text{m}$ is about a few cm/s. With this assumption, the sticking criterion is changed to the computation of the critical velocity. If the particles are slower than the critical velocity, then $H_{i,j} = 1$, else $H_{i,j} = 0$. Note, to incorporate the threshold diameter, $H_{i,j}$ is set to zero if $V_{i-j} + V_j > V_N$.

5.3 Bus stop model

For the computation of agglomeration, the population balance must be discretized and solved. It must be discretized into different particle diameter classes as well as over the entire 3D CFD domain. It would be simple to create an additional scalar field transport equation for each diameter in an Eulerian granular phase model. However, this approach requires considerable amounts of computational effort. Therefore another solution has to be found. The solution to this problem is to discretize the Smoluchowski [1917] PBE using the Lagrangian trajectories, which yields the Bus stop model. In this case, the system of equations will be solved directly by the Lagrangian simulation, and the agglomeration rates can be computed easily in the Lagrangian discription. Therefore, additional information from the Eulerian granluar phase at the specific tracer position will be needed. The idea of Bus stop model is shown in Figure 5.6, and can be described as follows. The particle trajectories are computed sequentially, beginning with the smallest particle class, d_{min}. If a particle, d_{i-j}, agglomerates with another particle d_j to form an aggregate of particles d_i it will stop at this position and wait for the aggregate of particles d_i.

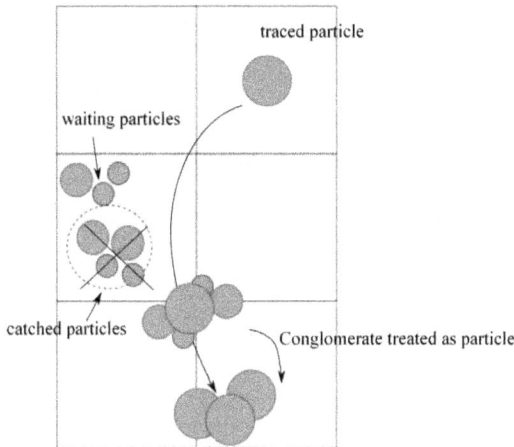

Figure 5.6: Bus stop model sequence: Particle cloud with diameter d_i is traced and finds two particles d_j and d_{i-j} which agglomerate to a particle d_i. These are added to the traced cloud.

Every particle trajectory checks if in each cell, smaller particles should be gained or lost. If particles should be gained, then it is checked if enough free particles of the relevant classes are waiting in the cell. This is done to avoid errors during the computation. Under normal conditions it would not be possible to get a number which is higher than the number of waiting particles. The waiting particles are collected by the trajectory and the particle number of the involved particle class is decreased in the cell. Hence, this does not change the diameter, nor the direction of the trajectory. Note, when calculating Lagrangian tracer particles in parallel with the Eulerian granular phase, the agglomeration rates can be computed directly in the Lagrangian description during a Lagrangian timestep. Additionally, the volume exchange between particle classes by agglomeration can also be included. The Bus stop model has only been developed for use with Lagrangian tracer trajectories, so it must be adapted to be valid for simulations with Lagrangian tracer particles.

5.3.1 Implementation

The Bus stop model is implemented in FLUENT. Therefore, some restrictions, determined by the software, must be considered, e.g. particle trajectories are not computed in parallel. This results in an implementation of the Bus stop model algorithm that is not optimised. Since we use the PBE, the proposed equations for the well observed agglomeration rates, $K_{i,j}$, can be used directly, which is in contrast to the Volume balance model (Chapter 5.4). The block diagram of the Bus stop model is sketched in Figure 5.7. It is divided into three consecutive steps: precomputations, the main part and post-correction of mass. In the first step, the tracer trajectories are computed without agglomeration modelling. In each cell which is hit by a trajectory, the particle velocity and particle number is stored. In the main part, the agglomeration rates, $K_{i,j} = K_{j,i}$, are calculated for each possible pair of particle collisions. If in a cell, a particle with diameter, $d_{p,i}$, is not present, the agglomeration rate for all pairs with this class is zero. Then the computation of tracer trajectories is repeated, while the amount of lost and gained particles at each position on each trajectory can be determined. Since the smallest particle class can only lose mass, this class is calculated first. In each cell, the lost particles are stored as waiting particles, which can be gained by other trajectories crossing this cell. After that, all other classes, in order of increasing diameter, are computed. During the trajectory computation, the mass (number of particles) of lost and gained particles is computed for each cell that is crossed by the trajectory. In the final step, particles which are not gained and are still waiting in one of the cells, were added to their particle trajectory again. With this restrictive procedure, the mass balance is always fulfilled as the mass of incoming and outgoing particles is the same. When considering the computational effort for the case of the preheating tower, this approach seems to be a good one. One disadvantage is that, in FLUENT, the trajectories are computed in serial and not in parallel. Therefore, the trajectories must be computed twice. If the trajectories are computed in parallel, the agglomeration rates could be accounted for after each time step. Then it is possible to compute the agglomeration during the tracer calculation. This would save computational time and effort.

5.4 Volume population balance model

Based on the idea of Smoluchowski [1917], and with respect to industrial applications, a volume population balance model (VPM) is presented here. The idea of this model is to use the volume of each particle class instead of the particle number and to include particle-particle collisions of more than two particles. Also, the high number of particle populations that can lead to numerical errors is avoided because of numerical limits. The requirements of the volume population balance model are given below

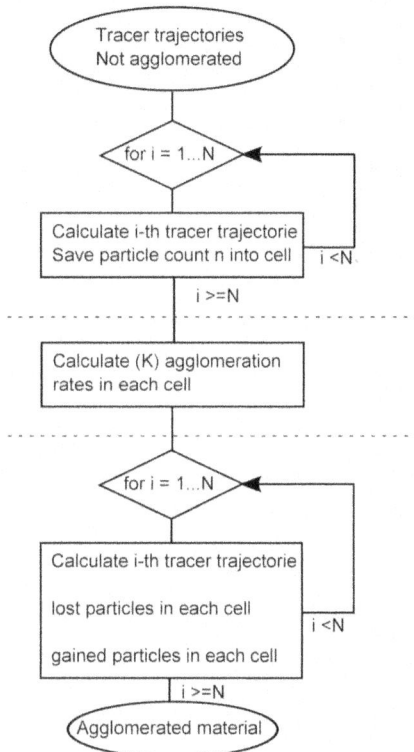

Figure 5.7: Block diagram of the Bus stop Model. The model consists of three main steps. First, the pre-computation step for information creation. Second, the computation of agglomeration rates. Thirdly, the simulation of agglomeration with post-correction of mass.

Agglomeration rules

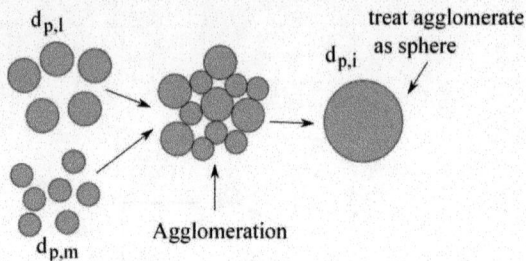

Figure 5.8: Simple sketch of the assumed agglomeration process by the volume population balance model

1. Agglomeration can not create or destroy particle mass $\sum\limits_{i=1}^{N} \frac{\partial h_i(t) V_{cell}}{\partial t} = 0$

2. Particle class l can only agglomerate with another particle classes m and create volume in a particle class i with a higher diameter, $d_l < d_i$ and $d_m < d_i$ and $d_l + d_m \leq d_i$

3. The biggest particle class considered, $i = N$ can not agglomerate

4. A probability value, $G_{i,l,m} = G_{i,m,l}$, denotes the particle volume production per second for particle class i, which is when particle class l agglomerates with m.

5. The agglomerated particles are treated as spheres.

6. The values of volume production per second, $G_{i,l,m}$, are assumed to be constants gained by experiments or estimations.

Following these rules the following system of equations was developed. For a particle class $i < N$, in case of N particle classes, the rate of change of the (i-th) particle class volume fraction, h_i, is given by

$$\frac{\partial h_i(t)}{\partial t} V_{\text{cell}} = -\frac{1}{2} h_i(t) \sum_{l=1}^{N-1} \left(\sum_{m=1}^{N} G_{i,l,m} \right) h_l(t) + \sum_{l=1}^{i-1} h_l(t) \left(\sum_{m=1}^{l} G_{i,l,m} h_m(t) \right)$$
$$- \frac{1}{2} h_i(t) \left(\sum_{m=1}^{i-1} G_{i,i,m} \right) h_i(t) \quad (5.41)$$

and for $i = N$ the equation changes to

$$\frac{\partial h_i(t)}{\partial t} V_{\text{cell}} = \sum_{l=1}^{i-1} h_l(t) \left(\sum_{m=1}^{l} G_{i,l,m} h_m(t) \right). \quad (5.42)$$

since, for particle class d_N, no particles can be lost by agglomeration. This model can be used if the population balance model will fail due to unknown physical behavior of the solid material. The biggest advantage of this model is that the particle diameter distribution can be freely chosen. In particular, the approach with fixed production rates will fail if exact modelling of the region where agglomeration takes place is necessary. In that case, a model based on particle distribution, particle-particle collisions and flow behavior has to be used, e.g. a model for the production rates $G_{i,l,m}$ has to be created. Another disadvantage is that for every new material, the volume creation rates, $G_{i,l,m}$, must be found by experiments and calibrated to fit to known results. For simulations that need to be solved quickly, and that are of plant-scale size, this model can be used to get an initial idea of the agglomeration behaviour of the industrial plant.

6

Validation by lab-scale experiments

The hybrid model, EUgran+Poly, and the implementation of the Bus stop agglomeration model are validated separately. Since the hybrid model was developed for industrial applications, the simulation setups are chosen with respect to common industrial simulation parameters. For example, the grid cell dimensions are bigger than in detailed academic simulations. Furthermore, all simulations are computed with a Courant number of $Cr = 0.95$ and an explicit scheme for the volume fraction. The solid phase is only computed and considered for values of $\alpha > 10^{-8}$.

The hybrid model, EUgran+Poly, is validated for two specific cases. First, the pneumatic conveying of poly-dispersed particles through a duct bend geometry at low mass loading is considered. The simulation is performed with different numerical models, and are compared to measurements given in Kuan et al. [2007]. Kuan et al. [2007] and Mohanarangam et al. [2007] showed nearly perfect simulation results for the gas and particle velocities for this problem. However, they used models which were specially developed for this duct bend and high grid resolutions. Therefore, we do not compare their results with ours, because of the different simulation setups. In the second case, a medium, mono-dispersed, particle-laden flow through a double-loop duct bend is studied. Here, the dispersion of a particle strand is compared to visual measurements. The measurements for the double-loop experiment were performed in the Johannes Kepler university laboratory and were presented in the work of Pirker et al. [2010]. the velocity and the distribution of fluid volume was compared to the measurements. Furthermore, a comparison of a pure Eulerian granular phase simulation and a EUgran+Poly hybrid model simulation is presented. In a third validation example. the implementation of the Bus stop agglomeration model is validated for a vertical pipe, where a poly-disperse granular medium is freely falling down. The change of the particle size distribution was compared with the results of the Smoluchowski equation after specific distances. In this case. the results of the Bus stop model are compared to simulation results with a direct discretized population balance equation. Thus. the implementation of the PBE was validated for this case only.

6.1 Dilute poly-dispersed flow in a duct

As mentioned before. the duct bend geometry of Mohanarangam et al. [2009] is used to validate the EUgran+Poly model, without agglomeration, for dilute poly-dispersed particulate conveying. This setup has been studied extensively by CFD simulations (e. g. Kuan et al. [2007] and Mohanarangam et al. [2009]).

Table 6.1: Simulation parameters used in the duct bend simulation.

Parameter	Value
u_{inlet}	$10 \, \mathrm{m/s}$
Re	about 10^5
L	0.00206
ρ_g	$1.18 \, \mathrm{kg/m^3}$
$\rho_s = \rho_p$	$2500 \, \mathrm{kg/m^3}$
$e_w = e_p$	0.9
μ_w	0.4
t_{ts}	$0.0005 \, \mathrm{s}$

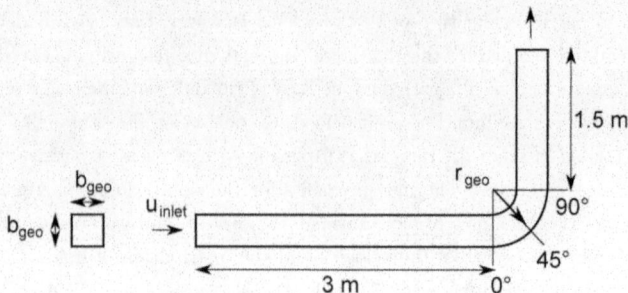

Figure 6.1: Geometry of duct bend [Kuan et al., 2007]

6.1.1 Boundary conditions and simulation setup

In Figure 6.1, the duct bend geometry is shown. The duct is made of steel, which is important for the boundary condition parameters, e. g. e_w and μ_w. The width of the rectangular cross-section is $b_{geo} = 0.15\,\mathrm{m}$ and the radius of the bend is $r_{geo} = 0.225\,\mathrm{m}$. Considering that coarse grids are needed for industrial systems, the geometry was meshed with 144000 cells, and 256 faces at the inlet. Starting from each face, one tracer trajectory is computed. The results revealed that this amount of trajectories was high enough to get a sufficiently accurate force exchange field. The gas and particle velocity at the inlet is set to $10\,\mathrm{m/s}$, implying a Reynolds number of about 10^5. The mass loading is $L = 0.00206$ and the density of the glass beads is $2500\,\mathrm{kg/m^3}$. The restitution coefficient between the wall and the glass particles, and between the particles themselves, was given by $e_w = e_p = 0.9$. The wall was assumed to be rough, with a restitution coefficient $e_w = 0.9$, a frictional coefficient $\mu_w = 0.4$, and a wall angle range of $20°$. For the boundary conditions, the models presented in Chapter 2 and Chapter 3 are used. The simulation setup parameters are given in Table 6.1. The particle diameter distribution of the poly-dispersed glass spheres is given in Table 6.2, which results in a mean diameter of $77\,\mu\mathrm{m}$.

6.1.2 Results and discussion

In Figures 6.2(a)-6.2(c), comparisons of the computed tangential velocities for the gas phase in the hybrid model, and the measurements of Kuan et al. [2007] in the duct bend, are shown. Here, $r = 0$ denotes the position of the outer wall in the bend and $r = 0.15$ the position of the inner wall in the duct bend. The figures

Table 6.2: Particle diameter distribution [Kuan et al., 2007]

Diameter in μm	Amount in %
5	$2.001 \cdot 10^{-5}$
18	$5.202 \cdot 10^{-5}$
30	$3.112 \cdot 10^{-3}$
45	$2.905 \cdot 10^{-1}$
57	3.26
65	10.35
76	38.10
89	34.03
103	9.982
125	2.5
140	1
152	$4.887 \cdot 10^{-1}$

reveal a fairly good agreement between the simulation results of the gas phase in the EUgran+Poly model and the measurements. Only at 90° are the results slightly different in the inner region. This can be attributed to the low grid resolution.

Figures 6.3(a)-6.3(c) show comparisons of the computed tangential velocities for the Lagrangian tracer particles and the measurements of [Kuan et al., 2007] in the duct bend. The simulation results for different numerical models and simulations are presented. At position 0° (Figure 6.3(a)), the differences between the mono-dispersed simulation, which used the average arithmetic particle diameter of Table 6.2 as the diameter, and the poly-disperse simulation appear negligible. However, it is observed that in the inner region ($r > 0.1$) no particles are found in case of mono-dispersed particles, since these are more affected by gravity than smaller particles in case of the poly-dispersed grains. Indeed, Figure 6.3(b) yields a considerable difference between mono-dispersed and poly-dispersed simulations, which is mainly caused by centrifugal forces. Regions near to the outer wall (i. e. $r = 0$) of the bend, appear not to be affected by neglecting the poly-disperse particle distribution. Figure 6.3(c) shows good agreement between the EUgran+Poly model and measured data, which is in contrast to the mono-dispersed results. In particular, the velocity magnitude is underestimated in the case of mono-dispersed particles (Figure 6.3(c)). Since the simulation was not able to reproduce the measured particle flow in the inner region ($r > 0.12$), none of the simulations could predict the particle velocities in this region correctly. This could be due to no stream detachment, resulting from the coarse grid resolution. It can be concluded that accounting for poly-dispersity in the numerical model has a big impact on the simulation results.

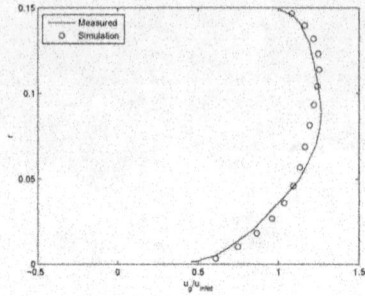

(a) Position 0°

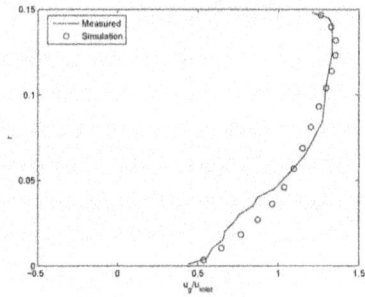

(b) Position 45°

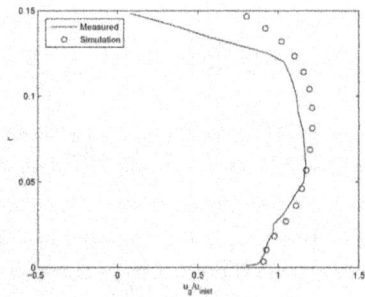

(c) Position 90°

Figure 6.2: Comparison of the tangential velocities of the gas phase for the hybrid model and measurements at three different positions in the bend.

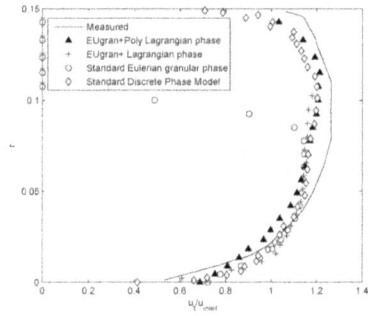

(a) Position 0°

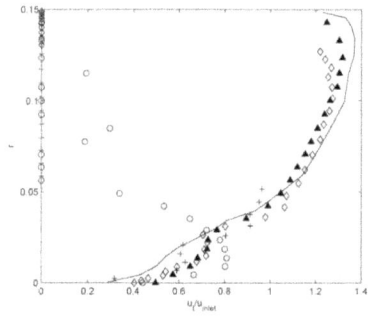

(b) Position 45°

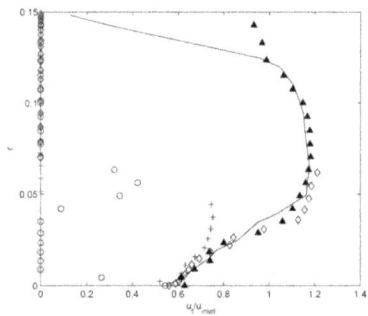

(c) Position 90°

Figure 6.3: Comparisons of the tangential velocities for the mono-dispersed Eulerian granular model and the poly-dispersed EUgran+Poly model at three different in the bend.

83

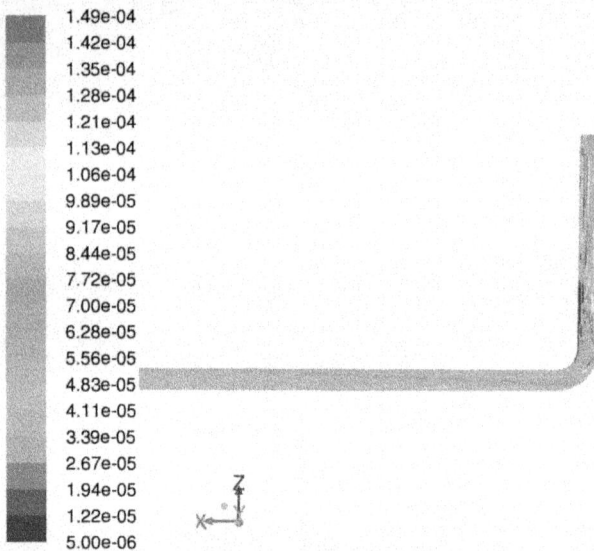

1.49e-04	
1.42e-04	
1.35e-04	
1.28e-04	
1.21e-04	
1.13e-04	
1.06e-04	
9.89e-05	
9.17e-05	
8.44e-05	
7.72e-05	
7.00e-05	
6.28e-05	
5.56e-05	
4.83e-05	
4.11e-05	
3.39e-05	
2.67e-05	
1.94e-05	
1.22e-05	
5.00e-06	

Figure 6.4: Particle size distribution in the bend. At the surface of the particle strand a bigger diameter as in the particle strand at the wall can be seen.

Figure 6.5: The wind tunnel in our academic laboratory. Most of our experiments are done at this facility.

Figure 6.4 shows the particle size distribution in the center plane of the duct bend. It can be observed that at the inner surface of the particle strand a higher average diameter is present as inside the particle strand at the wall. Two effects force this particle size distribution, bigger particles have a longer relaxation time as smaller ones and smaller particles are catched between wall and bigger particles. After the bend a dispersion of the strand and an averaging of the diameter distribution can be observed. It can be concluded that a significant difference to an uniform particle diameter distribution is present. However, currently there is no measurement for a validation of the particle diameter distribution in the duct available.

6.2 Mono-dispersed flow in a medium laden duct

After validation in low laden particle-laden flow regime the model behavior in medium regime is proven. In Figure 6.5, the wind tunnel for particle-laden flow experiments is shown. In this wind channel, the double-loop geometry (Figure 6.6) can be incorporated. Pirker et al. [2010] validated the mono-disperse hybrid model EUgran+ on this setup for medium particle-laden flow, which also provided a huge amount of validation data [Goniva et al., 2012, Kloss, 2011, Kahrimanovic, 2009].

Figure 6.6: Picture of the double-loop geometry. The double-loop geometry was installed in the wind tunnel and used for several experiments in our laboratory.

In the double-loop geometry (Figure 6.6), the particle-laden flow is forced by centrifugal forces to form a particle strand. After the second loop, the dispersion of particle strands can be observed. The behavior of the strand dispersion is simulated numerically, and visually validated by experiments. Pirker et al. [2010] showed that the standard Lagrangian discrete phase model and the standard Eulerian granular phase model fail to correctly predict the dispersion of the particle strand.

6.2.1 Boundary conditions and simulation set up

In Figure 6.7, the double-loop geometry is shown. At the inlet, the double-loop is fed with mono-dispersed particles of diameter $d_p = 0.084$ m. In contrast to Pirker et al. [2010], we used the viscosity modelling of Hrenya and Sinclair [1997] and the boundary conditions of Schneiderbauer et al. [2012b] instead of the viscosity model by Syamlal [1987] and the boundary conditions of Johnson and Jackson [1987]. The results of the EUgran+Poly hybrid model are compared to velocity measurements and a pure Eulerian granular phase (EUgran) simulation at three positions. In addition, the volume fraction of the granular medium is measured at Pos. 1 and Pos. 2 (shown in Figure 6.7) and compared to the simulation results.

In Table 6.3, the simulation parameters for the double-loop simulation are given. It can be seen that the particle diameter ($d_p = 0.084$ m) is about 1000 times larger than in the duct bend case. Additionally, the mass loading is on average $L = 1.5$. This, in turn, implies that a different regime of particulate transport must be considered in this case. The particle mass-flow distribution at the inlet is linearly interpolated, as in the experiment, to get a good agreement between measurement and simulation.

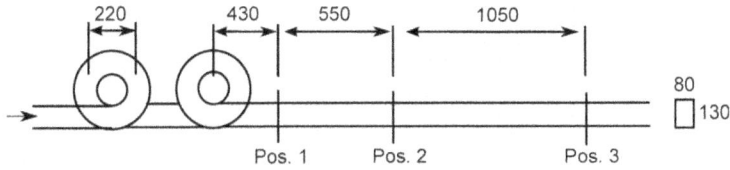

Figure 6.7: Double-loop geometry with velocity measurement positions [Pirker et al., 2010].

Table 6.3: Simulation parameters for double-loop geometry

Parameter	Value
u_{inlet}	$18.5\,\text{m/s}$
L	1.5
ρ_{g}	$1\,\text{kg/m}^3$
$\rho_{\text{s}} = \rho_{\text{p}}$	$2500\,\text{kg/m}^3$
e_{p}	0.9
e_{wall}	0.91

In addition, due to the circular flow of air in the wind tunnel, it was observed that the temperature in the wind channel rises from about $20°$ of Celsius up to $60°$ of Celsius. This changed the density of air, which was accounted for in the simulations by adapting the density of air (Table 6.3). For the particle-wall handling the parameters of the duct bend setup are used.

6.2.2 Results and discussion

In Figure 6.8, a comparison of the measured and computed particle strand at the end of the second loop is shown. It was observed that the dispersion of the strand is in good agreement with the experiment. A visual comparison revealed that the measured and computed dispersion rate were in the same range. Compared to the results obtained by standard models presented in Pirker et al. [2010], the modifications of the Eulerian granular phase show a significant improvement in the simulation results. In the next validation step, it was quantitatively evaluated whether the new models used for the Eulerian granular phase provided acceptable results. Figure 6.9 shows a comparison between measurements for a pure Eulerian granular phase simulation, and the EUgran+Poly simulation for velocity and particle concentration. Figure 6.9 also illustrates that the computed particle velocities for the cases of the modified pure EUgran and EUgran+Poly model correlate well with measured data.

(a) Dispersion of particle strand in the experiment.

(b) Result of a pure Eulerian granular phase model simulation using the new models.

Figure 6.8: Comparison of experiment and simulation. Both show the dispersion of the particle strand behind the second loop.

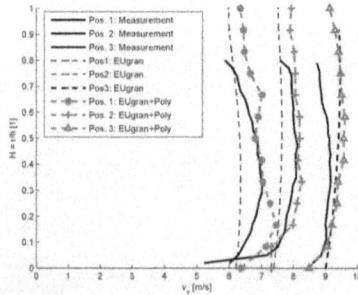

Figure 6.9: Average particle velocity at position 1, 2 and 3 after the second loop.

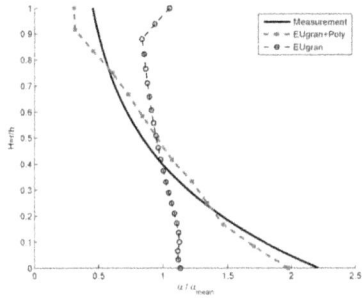

(a) Volume fraction at position 1

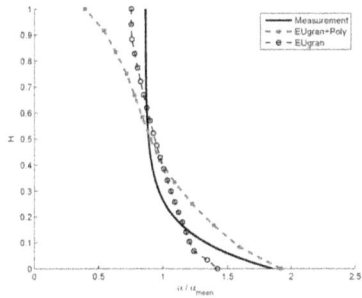

(b) Volume fraction at position 2

Figure 6.10: Comparison of measured and simulated particle parameters. Volume fractions were compared at two positions behind the second loop.

This was also the case for the mono-dispersed hybrid model of EUgran+ [Pirker et al., 2010] (not shown in the figure). In Figures 6.10(a) and 6.10(b), the volume fraction in the channel at two positions is shown. These figures reveal a considerable influence of the Lagrangian discrete phase in the hybrid model. Thus, when compared to the pure Eulerian granular phase model, the hybrid model results are significantly improved. This is attributed to the Magnus force [Sommerfeld and Huber, 1999, Kahrimanovic, 2009] and the additional rough wall boundary condition model, which are both accounted for the Lagrangian discrete phase. The validation results indicate that the hybrid model even improves the simulation results in mono-dispersed particle-laden flows. To conclude, the EUgran+Poly can be used for the simulation of mono- and poly-dispersed, low and high mass, particle-laden flows for pneumatic conveying.

6.3 Agglomeration of poly-dispersed particulate flow in a vertical pipe

For the validation of the Bus stop model a simple geometry was chosen. This geometry, which is a vertical pipe, is shown in Figure 6.11. As a first validation example, limestone particles with a density of $\rho_p = 2700 \, \text{kg/m}^3$ are falling freely in a pipe of length $l = 50 \, \text{m}$ and square diameter $d = 0.1 \, \text{m}$ with a terminal settling velocity. According to equation (6.1), 20 particle classes were used, with $d_1 = 5 \, \mu\text{m}$. The threshold for the agglomeration was set to $d_p = 92.94 \, \mu\text{m}$ in this case. The particle size distribution was chosen to be Gaussian

$$n_i = C_1 e^{\left(-0.5\left(\frac{d_i - d_c}{d_\sigma}\right)^2\right)} \tag{6.1}$$

with $d_c = 8.75 \, \mu\text{m}$ and $d_\sigma = 1 \, \mu\text{m}$. For the given volume fraction of particles, $\alpha_p = 0.3$, C_1 can be calculated. In a second validation example, for the same geometry, another particle size distribution with $d_1 = 1 \, \mu\text{m}$, $d_c = 1.75 \, \mu\text{m}$ and $d_\sigma = 0.2 \, \mu\text{m}$ was used. The Smoluchowski model simulation in MATLAB was stopped after 99% of mass reached the end of the pipe, otherwise the computation would have taken to much time due to very slow convergence of the last 1%. Thus a small error was present in the mass conservation, which was neglected in the validation. Figure 6.12 shows a comparison of the particle size distribution at the end of the 50 m pipe between the Bus stop model and the Smoluchowski population balance model.

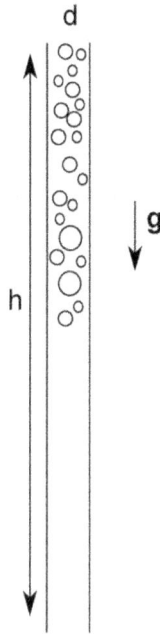

Figure 6.11: Sketch of the vertical pipe, which was used to validate the agglomeration model

It can be seen that both models show nearly identical results. The comparison between the Bus stop model in MATLAB and FLUENT shows minor differences. These negligible differences were caused by setting the velocity of particles in the MATLAB model to the fixed particle settling velocity, whereas the particle velocity in FLUENT was computed based on the drag. This yielded an error of about $10^{-4}\,\mathrm{m/s}$, which was neglected. The result of the second particle size distribution is given in Figure 6.13, which illustrates the change in particle size distribution within the pipe from 1 m to 50 m. The results show the same quality as for the first validation case. Furthermore, the results show that the Bus stop model can be used for the simulation of agglomeration with Lagrangian tracer trajectories, and that it should be accurate enough for the simulation of centrifugal dust separators. To conclude, the use of the Bus stop model is justifiable.

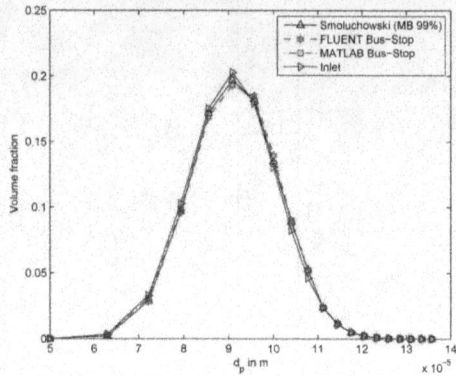

Figure 6.12: Comparison of particle volume fraction between discretized Smoluchowski model and the Bus stop model (in MATLAB and FLUENT)

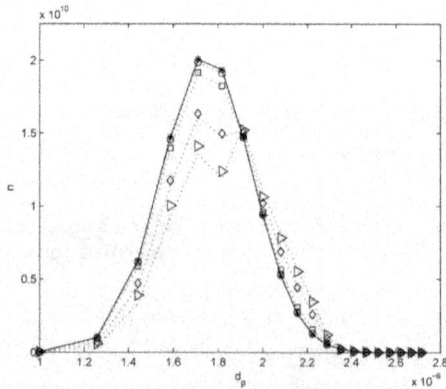

Figure 6.13: Change of particle size distribution with height ($\bigcirc$ 1 m, $\square$ 5 m, $\diamond$ 25 m and $\triangleright$ 50 m) in the vertical pipe. Comparison between Smoluchowski and Bus stop model.

7

Application to cyclone separation

The validation results for dilute and medium granular flows were satisfying. Hence, the EUgran+Poly hybrid model and the Bus stop model were used for the simulation of a cyclone to show the applicability of these models. Therefore, two industrial dust separation cyclones, with a poly-dispersed granular flow in high and low mass loading conditions were simulated. The simulation of the experimental-scale cyclone was compared to an analytic calculation based on Muschelknautz et al. [1994]. For the theory of cyclone modelling the reader is referred to the book of Hoffmann and Stein [2008]. Additionally, a comparison of the computational efficiency between an Eulerian granular phase approach with N particle phases, and the EUgran+Poly hybrid model is shown. For a second application, an industrial cyclone was studied, and the simulation results were compared to measurements. Therefore, the separation efficiency of different particle diameter classes was studied. Lastly, the impact of the agglomeration model was studied and discussed for a simple cyclone simulation, but without any validation based on measurements.

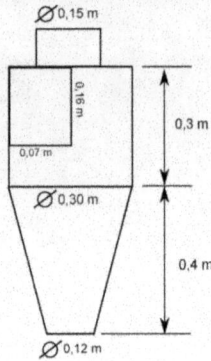

Figure 7.1: Geometry of the in-house cyclone.

7.1 Hybrid Model

7.1.1 Boundary conditions and simulation setup

Two different cyclones were used for the simulations. One was our in-house cyclone, and the other was an experiment-scale, industrial cyclone, used for the separation of limestone. Theoretical and measured data is available for both cyclones.

In-house cyclone

The cyclone geometry is illustrated in Figure 7.1, and a picture of the cyclone is shown in Figure 7.2. The cyclone was designed with the theory of Muschelknautz et al. [1994] and Hoffmann and Stein [2008] to separate particles of $d_p > 10^{-6}$ m with an efficiency of 99 % by an inlet velocity of 25 m/s. A cyclone vortex Reynolds number of $Re_g = 7120$ was calculated for this case with an inlet speed of 25 m/s. In his doctoral thesis, Kahrimanovic [2009], simulated the cyclone with the Lagrangian discrete phase model, which included statistical particle collisions based on the model of Tsuji et al. [1989]. EUgran+Poly model simulations, with a poly-dispersed particle loading of $L = 0.01$ were performed and compared to the Muschelknautz theory.

A second simulation was computed with mass loading of $L = 0.095$, and an inlet velocity of 15.6 m/s. For this setup validated measurement of the pressure drop are available. To account for the gas temperature change, because of the circu-

Figure 7.2: Picture of the in-house cyclone.

lar flow, the air density must be changed to $\rho_g = 1.114\,\text{kg/m}^3$. In Table 7.1, the simulation parameters for the cyclone simulation are given. The particles used for

Table 7.1: Simulation parameters for the in-house cyclone

Parameter	Value
u_{inlet}	$25\,\text{m/s}$
L	0.01
ρ_g	$1\,\text{kg/m}^3$
$\rho_s = \rho_p$	$800\,\text{kg/m}^3$

measurements and simulations in our experiment-scale cyclone were hollow glass spheres. Glass has a density of about $2500\,\text{kg/m}^3$, but the spheres were hollow and therefore the density was reduced to $800\,\text{kg/m}^3$. The diameter distribution of the poly-dispersed granular glass phase is given in Table 7.2.

Industrial limestone separation cyclone

In a second validation example, different numerical models were compared with measurements of an industrial cyclone used for the separation of limestone (Figure 7.3). The dimensions of the rectangular entrance of the cyclone are bigger than our in-house cyclone. The mass loading is $L = 1$. In Table 7.3, the simulation parameters for the cyclone simulation are given.

The particle diameter distribution for this case is shown in Table 7.4.

95

Table 7.2: Particle diameter distribution for glass particle

Diameter in μm	Amount in %
1	0.0023
2	0.03
4	0.5
8	3.34
12	10.1
16	19.6
20	44.5
30	21.9

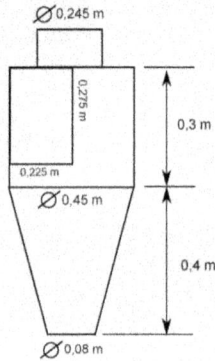

Figure 7.3: Geometry of the limestone separation cyclone.

7.1.2 Results and discussion

In-house cyclone

The efficiencies for low and high mass loading are nearly identical, because particles with a diameter much higher than the cross-grain diameter of the cyclone were used in the simulations. Thus, nearly all particles were deposited. Figure 7.4 shows the separation efficiency of the in-house cyclone compared to the Muschelknautz theory for a mass loading, $L = 0.01$. The theoretical curve and the simulation results agree very well. Furthermore, the separation efficiency of the cyclone in the simulations was between $98, 2 - 99, 7$ percent. The estimated separation efficiency by Muschelknautz theory was 99.7 percent for the cyclone. To gain further information on the pressure loss, another simulation was done. For this case, the inlet velocity was set to $v_{in} = 15.6\,\text{m/s}$, because pressure measurements were available for this

Table 7.3: Simulation parameters for the limestone separation cyclone

Parameter	Value
u_{inlet}	$10\,\text{m/s}$
L	1
ρ_g	$1\,\text{kg/m}^3$
$\rho_s = \rho_p$	$2700\,\text{kg/m}^3$

Table 7.4: Particle diameter distribution for limestone particles

Diameter in μm	Amount in %
0.55	3.58
0.9	6.29
1.55	11.74
2.5	12.57
4.5	14.68
7.5	11.54
1.25	10.35
2.15	10.09
3.65	10.08
6.15	9.08

inlet velocity. The pressure loss in the cyclone simulation was estimated to be 306 Pa (mixture) to 440 Pa (only gas). In comparison, the measurement shows a pressure loss of 360 Pa. To reach the value given by measurements, the inlet speed for the gas had to be increased to 17.9 m/s in the simulation. On one hand this difference could be reduced by adapting the wall treatment to smoother walls, but on the other hand the result gives a good first estimate of the range of pressure loss inside the cyclone. In Figure 7.6, the particle strand which appears inside the cyclone can be seen. This strand is forced by the poly-disperse drag law which includes the shadow effect in particle strands. In Figure 7.5, two cross sections of the volume fraction in the middle plane of the cyclone, is shown. Different color scales are used for Figure 7.5 and Figure 7.6, so it seems like different cases were used. The different scaling was chosen to give a better overview of the volume fraction in Figure 7.5.

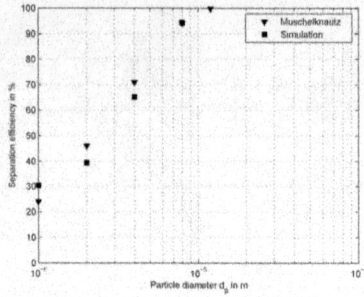

Figure 7.4: Separation efficiency compared to Muschelknautz theory for a mass loading $L = 0.01$

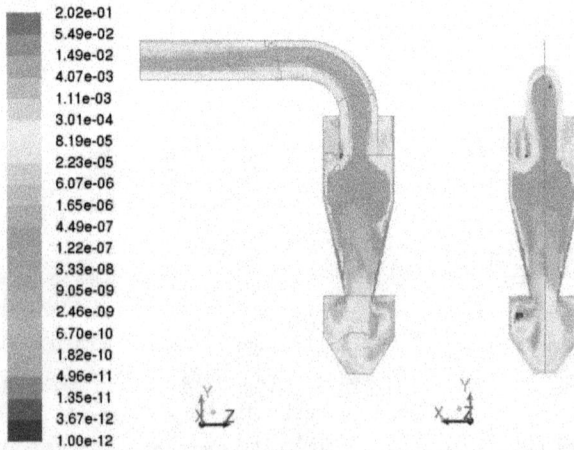

Figure 7.5: Volume fraction in the cyclone. The volume fraction is scaled logarithmically.

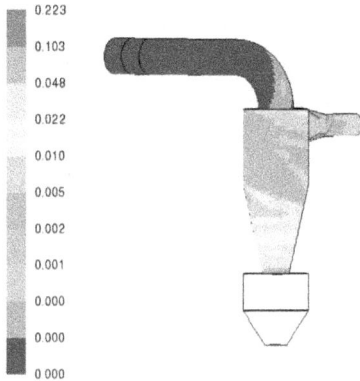

Figure 7.6: Volume fraction, α_s, in wall adjacent cells for $L = 1$. The particle strand formation can be seen.

7.1.3 Results and discussion for separation of limestone material

In Figure 7.7, the separation efficiency of the cyclone for different numerical models is compared to measured results. It can be seen for the case of the discrete phase model, where the interaction between particles was neglected, the small particles were not trapped by the strand in the cyclone, and so they were not separated. This is also the case if ten individual Eulerian granular phase simulations, one for each particle diameter, are computed. Thus it can be assumed, for the given mass loading, neglecting particle collisions gives the wrong simulation results. The results of the hybrid model dramatically shows the impact of the particle-particle collisions and the particle drag force for poly-dispersed particulate flows. A coupled simulation with ten Eulerian granular phases, was too time consuming (see Chapter 7.1.4) and was therefore not performed. Both curves of the EUgran+Poly Model show two different simulation setups with respect to the particle inlet. Obviously, in industrial facilities, the particle inlet is optimized to reduce pressure loss and to place material where it is needed. Additionally, a compact particle inlet design encourages particle strand formation and achieves a higher separation efficiency. In the worst case, the material is homogenously distributed at the inlet, which is shown by the homogeneous curve in Figure 7.7. In the other case, the particle distribution at the inlet is fitted to achieve the particle distribution of the real experimental setup. This results in a heterogenous particle distribution at the inlet of the simulation

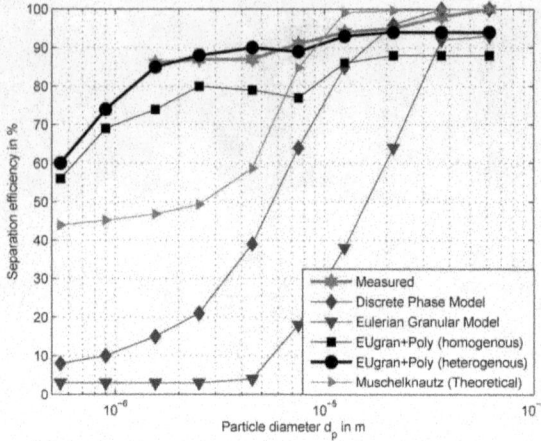

Figure 7.7: Comparison of separation efficiency for different numerical models for mass loading $L = 1$

geometry. The improvement of the simulation results shows the importance of the correct modelling of particle feeding.

7.1.4 Discussion of computational efficiency

To show the impact of computational effort reduction, both a hybrid model, and a simulation with ten Eulerian granular phases, were implemented. In Table 7.5, the comparison between the Eulerian granular phase simulation and the EUgran+Poly model can be found. It can be seen that the computational time is about ten times higher than the time required for the hybrid model simulation. This is the case because for the Eulerian granular system, ten phase equation systems must be solved, but for the hybrid system, only one needs to be solved. However, for the hybrid approach, K tracer particles are computed every N time steps, which ends up in K uncoupled differential equations. All-in-all, the hybrid approach requires less equations to be solved, because the K equations must only solved for differential time steps and not in space.

Table 7.5: Comparison of computational effort, averaged over 1000 time steps

Parameter	Eulerian Phases	EUgran+Poly	Reduction
cpu time	9.224 s	0.737 s	> 10
Data transfer	51.6 MB	14.9 MB	> 3.5

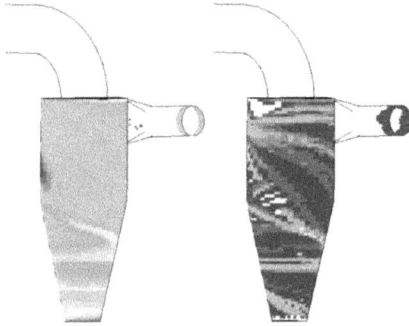

Figure 7.8: Volume fraction of granular material in the cyclone. On the left side, the Eulerian-Eulerian granular phase and on the right side the coupled Lagrangian tracer trajectories can be seen.

7.2 Agglomeration

The in-house cyclone was used to test the agglomeration model on an industrial-scale application. Geometry and results for the cyclone, without agglomeration modelling, can be found in Section 7.1. The same particle size distribution as mentioned before in the validation chapter (equation 6.1 and 20 particle classes with $d_1 = 5\,\mu\text{m}$) were used for the simulation. Figure 7.8 shows the volume fraction in the granular phase and the amount of particles in the Lagrangian trajectories. In both representations, the same particle strand can be seen. The change in particle size distribution by agglomeration, for each particle class, can be found in Figure 7.9. The change in volume rates is small, so the results must be used very carefully. Nevertheless, the movement of smaller particles to bigger ones by agglomeration can be observed. In doing so, the agglomeration model increases average diameter in the cyclone, and therefore the separation efficiency in the simulation. The smaller particles agglomerate with other particles to form particle aggregates which are in a diameter range that is separated. As a result, a better simulation of the real effects in a cyclone is possible by taking agglomeration into account. In this example the agglomeration model increased the separation efficiency of the simulation by 1-4 %.

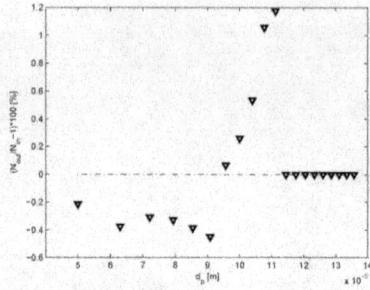

Figure 7.9: Difference of particle size distribution at the inlet and at the particle outlet of the cyclone.

However, since no measurements of the increase of separation efficiency are available for the presented cyclone and material, the results can not be compared to a real world application. for given simulations, without agglomeration modelling, the agglomeration model can be used to give a rough estimate of the particle diameter distribution at the outlet, also receiving a rough estimation of the separation efficiency with agglomeration. However, an EUgran+Poly simulation, with the agglomeration model, will provide a different result because of the changing diameter in the Eulerian granular phase.

If man made himself the first
object of study, he would see how
incapable he is of going further.
How can a part know the whole?

Blaise Pascal (1623-1662)

8

Conclusions and Outlook

In this book, a hybrid model for the simulation of particle-laden, poly-dispersed flows was presented. The model was an expansion of the mono-dispersed hybrid model EUgran+ presented by Pirker et al. [2010]. A novel implementation of the population balance equation by Smoluchowski [1917], based on Lagrangian tracer trajectories, and referred to as the Bus stop model was also presented. The hybrid model was validated by experiments, and its application to an industrial cyclone was shown. Furthermore, it was demonstrated that the usage of the EUgran+Poly model reduces computational costs for simulations on an industrial scale.

Hybrid model: EUgran+Poly

In general, the hybrid model is a good choice for the simulation of particulate convey-ing and separation in dilute-to-dense granular flow regimes. The idea of computing the physical behaviour in the description which requires less computational effort, leads ro significantly better results when compared to more common models. The hybrid model is implemented in a modular way. Thus, the user can choose which models are active in the simulation, without the necessity of rewriting the code. In addition, it is possible to change models or expand the hybrid model with little effort. It was shown that the model can be used for the simulation of particle conveying, even for very small diameters in the range of μm. This is important for modelling industrial dust separation and transport systems. The main findings of this work are:

- **EUgran+Poly hybrid model**
  An Eulerian granular phase model, augmented with Lagrangian models provides, good results for particulate flows with dense and dilute mass loading regimes.

- **Interphase momentum exchange**
  A model of the poly-disperse particle drag force that recognizes the lee field of particles is both necessary and crucial. Standard drag models will fail in regions with volume fraction $\alpha > 10^{-5}$.

- **Wall treatment**
  The new derived boundary conditions by Schneiderbauer et al. [2012b], which distinguish between sticking and sliding particle-wall contact, improves the description, and accuracy, of the particle-wall interaction. It seems that the boundary conditions of Johnson and Jackson [1987] can now be replaced by the new approaches.

Agglomeration: Bus stop model

In this work, a new implementation of the particle population balance equation was presented, namely, the Bus stop model. It was shown that the Bus stop model solves the population balance on the Lagrangian-tracer trajectories. Thus, it gives the same results as the common solver for the PBE. It was also shown that agglomeration affects the average particle diameter in the cyclone, and therefore a higher separation efficiency was computed.

Outlook

In further research, the agglomeration model may be extended to support the break-up of agglomerates. A modelling of de-agglomeration would lead to the possibility of getting a stable particle size distribution behavior, which is reached at equilibrium between agglomeration and break-up. Additionally, the impact of external forces on particles could be included, for example electric or magnetic fields. With respect to software license costs, the implementation of the hybrid model in OpenFOAM is planned. Furthermore, models for heat exchange and chemical reactions can be included, e.g. for the simulation of calciners. Chemical reactions could also be included in further research.

Finally, the EUgran+poly and the Bus stop approach are efficient approaches for modelling particle-laden flow, especially for the simulation of particulate transport and separation systems, which are often found in industry.

A

Restitution coefficients are no constants

During the work on this book, two publications which expose the theory of restitution coefficients were considered. One from Pöschel, T. [2000], and the other from [Abedi, 2009]. During the introduction in the boundary conditions of Johnson and Jackson [1987] and the arbitrarly chosen specularity coefficient in the model, we thought more and more of a specularity coefficient independent model. This was found in the model of Jenkins and Louge [1997], which is based on Coulomb's law that decides between sliding and sticking collisions between particles and walls. Also, the model of Schneiderbauer et al. [2012b], which was developed during this period of the work, depends on Coulomb's law. But all these models need restitution coefficients in the normal and tangential directions to the wall. These describe the ratio of lost energy during the impact, and are commonly treated as constant. But Pöschel, T. [2000], in his work, describes the calculation of a restitution coefficient for the normal and tangential directions. In his model, the restitution coefficients depend on many properties, which will take too long to describe here. The measurements and results of Abedi [2009] were used. Abedi [2009] wrote that the restitution coefficients depend mainly on the impact velocity, the impact angle and the particle diameter at the impact position. The normal restitution coefficient, e_n, depends on impact angle and impact velocity, while the tangential restitution coefficient, e_t, on impact angle and particle diameter. The equations presented by Abedi [2009] were implemented in the hybrid model and tested in the low particle-laden duct bend case. The results were nearly the same as when the constant restitution coefficients were used. Therefore, no further development was done in this field, and the computation of the restitution coefficient was excluded from the hybrid model.

B

Computation of Lagrangian particle-wall collision

The algorithm used for the computation of the particle-wall collision in this book follows that of Kahrimanovic [2009], and is presented on the following pages.

1. **Calculate the normal vector of the wall**
   The normal vector, $\mathbf{n}_w$, of the wall is computed.

2. **Calculate impact angle of the particle**
   The impact angle, α, is

   $$\alpha = \frac{\pi}{2} - \cos\left(\frac{\mathbf{n}_w \cdot \mathbf{u}}{|\mathbf{u}|}\right). \tag{B.1}$$

3. **Calculate the restitution coefficient**
   α_e is chosen as the threshold value, $\alpha_e = 30°$. Note that the alpha angle

is normally in radians. $e_\mathrm{h} = 0.9$ is chosen as the maximum value for the restitution coefficient.

$$\alpha <= \alpha_c : e = (e_\mathrm{h} - 1)\frac{\alpha}{\alpha_c} + 1 \tag{B.2}$$

$$\alpha > \alpha_c : e = e_\mathrm{h} \tag{B.3}$$

4. **Calculate the friction coefficient**

The calculation of the friction coefficient is done with the same assumption, as the calculation of the restitution coefficient. α_μ is chosen as threshold value. $\alpha_e = 30°$, but in radians. $\mu = 0.3$ is the maximum restitution coefficient, and $\mu_0 = 0.4$

$$\alpha <= \alpha_\mu : f_c = (\mu - \mu_0)\frac{\alpha}{\alpha_\mu} + 1 \tag{B.4}$$

$$\alpha > \alpha_\mu : f_c = \mu \tag{B.5}$$

5. **Create a co-ordinate system at the collision point**

$$\mathbf{p} = \mathbf{v} \times \mathbf{n}_\mathrm{w} \tag{B.6}$$

$$\mathbf{x}_C = \frac{\mathbf{p}}{|\mathbf{u}|} \tag{B.7}$$

$$\mathbf{y}_C = \mathbf{n}_\mathrm{w} \times x_C \tag{B.8}$$

6. **Get stochastic values for virtual angles**

In three dimensions, two virtual angles are needed to simulate a rough wall [Kahrimanovic, 2009]. The angles α_virt and β_virt are chosen with a random number generator based on a Gaussian distribution. α_virt is taken to be between 0 and α, while β_virt between 0 and $90°$. It must be remembered that $90°$ is just the threshold, and will not appear in reality.

7. **Create rotation matrices from flat to rough wall co-ordinate systems**

The local co-ordinate system is directly on the wall and the normal vector is

$\mathbf{n}_l$ as shown in Fig. 3.7. The transformation is done by two transformations of the original co-ordinate system, $\mathbf{A}_{IC}$, and second a rotation by β_{virt} with $\mathbf{A}_\beta$.

$$\mathbf{A}_{IC} = \begin{pmatrix} x_{C,1} & y_{C,1} & n_{C,1} \\ x_{C,2} & y_{C,2} & n_{C,2} \\ x_{C,3} & y_{C,3} & n_{C,3} \end{pmatrix} \tag{B.9}$$

First, a rotation by α_{virt} with $\mathbf{A}_\alpha$

$$\mathbf{A}_\alpha = \begin{pmatrix} 1 & 0 & 0 \\ 0 & \cos(\alpha_{virt}) & \sin(\alpha_{virt}) \\ 0 & -\sin(\alpha_{virt}) & \cos(\alpha_{virt}) \end{pmatrix} \tag{B.10}$$

and second, a rotation by β_{virt} with $\mathbf{A}_\beta$

$$\mathbf{A}_\beta = \begin{pmatrix} \cos(\beta_{virt}) & 0 & \sin(\beta_{virt}) \\ 0 & 1 & 0 \\ -\sin(\beta_{virt}) & 0 & \cos(\beta_{virt}) \end{pmatrix}. \tag{B.11}$$

The full transformation matrix is given by the multiplication of the three matrices

$$\mathbf{A} = \mathbf{A}_{IC} \mathbf{A}_\alpha \mathbf{A}_\beta. \tag{B.12}$$

8. **Transform omega and velocity into the local co-ordinate system**

$$\mathbf{u}_{CS} = \mathbf{A}^T \mathbf{u} \tag{B.13}$$
$$\omega_{CS} = \mathbf{A}^T \omega \tag{B.14}$$

9. **Calculate velocity and rotation after collision**

There are two possible types of collisions, the sliding, and the non-sliding collision. For the decision as to which collision occurs, some calculations are necessary. First, the z-coordinate of the velocity ($\mathbf{v}_{AI}$) and the rotation speed (ω_{AI}) are computed with

$$u_{AR,3} = -eu_{CS,3} \tag{B.15}$$
$$\omega_{AR,3} = \omega_{CS,3}. \tag{B.16}$$

The index AR means "after reflection". Additionally, the velocity at the impact point must be considered.

$$u_x = u_{CS,1} + \frac{d}{2}\omega_{CS,2} \tag{B.17}$$

$$u_y = u_{CS,2} - \frac{d}{2}\omega_{CS,1} \tag{B.18}$$

$$u_R = \sqrt{u_x^2 + u_y^2} \tag{B.19}$$

A non-sliding collision is found if

$$u_R <= \frac{7}{2}\mu_0\left(1 + e\right)u_{CS,3} \tag{B.20}$$

- **Non-sliding collision**

$$u_{AR,1} = \frac{5}{7}\left(u_{CS,1} - \frac{2}{5}\omega_{CS,2}\right) \tag{B.21}$$

$$u_{AR,2} = \frac{5}{7}\left(u_{CS,2} - \frac{2}{5}\omega_{CS,1}\right) \tag{B.22}$$

$$\omega_{AR,1} = \frac{2u_{AR,2}}{d} \tag{B.23}$$

$$\omega_{AR,2} = -\frac{2u_{AR,1}}{d} \tag{B.24}$$

- **Sliding collision**

$$u_{AR,1} = u_{CS,1} - \frac{u_x}{u_R}f_c\left(e + 1\right)u_{CS,3} \tag{B.25}$$

$$u_{AR,2} = u_{CS,2} - \frac{u_y}{u_R}f_c\left(e + 1\right)u_{CS,3}\omega_{AR,1} = \omega_{CS,1} + \frac{5}{d}\frac{u_y}{u_R}f_c\left(e + 1\right)u_{CS,3} \tag{B.26}$$

$$\omega_{AR,2} = \omega_{CS,2} - \frac{5}{d}\frac{u_x}{u_R}f_c\left(e + 1\right)u_{CS,3} \tag{B.27}$$

10. **Transform vectors back to global system**

The new translational and rotational speeds are now known, and are transformed back into the normal co-ordinate system.

$$\mathbf{u} = \mathbf{A}\mathbf{u}_{AR} \tag{B.28}$$

$$\omega = \mathbf{A}\omega_{AR} \tag{B.29}$$

C

UDF Structure of hybrid model

The hybrid model consists of three main files (Fig. C.1): **activeModules.h**, **variables.h**, **EUgran_plus_Agglo.c**. In Figure C.1, the interaction between the files is shown. The simulation sequence is controlled by the file **calc.scm**. **activeModules.h** consists of boolean variables that define which numerical models in the hybrid model are active. This includes, for example, the activation of the Magnus force, granular pressure on tracer particles, coupling of Lagrangian and Eulerian phase, and agglomeration. A value of 1.0 activates the model, while a 0.0 de-activates the model. Some models depend on other models and can only be active if the other one is activated, e. g. the Eulerian Magnus force can only be used if the Magnus force is calculated for the Lagrangian tracer particles. In **variables.h**, all global variables which are needed by the hybrid model, have to be defined. Additionally, some of them must be defined in FLUENT again. For easier understanding of the hybrid model, this file summarizes all important variables for the UDF. For example, this includes the density, average particle diameter, particle mass flow rate and restitution coefficients. The file **EUgran_plus_Agglo.c** is the main file of the hybrid model, and contains the full C-code for the EUgran+Poly hybrid model and the agglomeration part. The

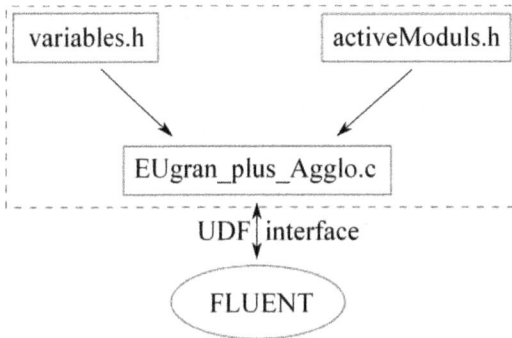

```
┌─────────────────────────────────────────────────────────┐
│  variables.h              activeModuls.h                │
│          ╲                    ╱                          │
│           ╲                  ╱                           │
│            ▼                ▼                            │
│          EUgran_plus_Agglo.c                             │
└─────────────────────────────────────────────────────────┘
           UDF │ interface
               ▼
           ( FLUENT )
```

Figure C.1: Connection between files and FLUENT.

calc.scm file is a SCHEME file for FLUENT. It describes the simulation procedure and guarantees that the simulation can run on its own. Thus no input from the user is required. The file consists of two simulation procedures, one with agglomeration and one without. In this file the number of time-steps between each computation of tracer-trajectories has to be chosen. Also, the number of complete cycles of the hybrid simulation cycles can be chosen. Considering Figure 4.6, the steps to use the EUgran+Poly hybrid model for the simulation of a granular flow are described on the following pages.

1. **Setup and initialization of the computational system**

   Before the simulation can be started, the simulation case must be set up. Therefore, the Lagrangian case, the Eulerian granular phase, and the coupling must be initialized.

   - **Setup of variables**

     In a first step, the properties of the granular should be written into the "variables.h" file.

   - **Setup of modules**

     In the "activeModules.h" file, the numerical models which are used by the simulation must be activated.

   - **Setup of particle size distribution**

     In the header of "EUgran_plus_Agglo.c", the particle size distribution

113

must be set correctly. Since, FLUENT only allows the usage of 500 user-defined memories, the number of particle classes is limited to 20 of them.

- **General setup**
  The mesh of the geometry must be loaded, reordered and scaled. If gravity is required, then it must be activated. The simulation must be computed as "time transient". The UDF has to be compiled and included into the simulation, otherwise the UDF methods cannot be included into the next step.

- **Setup of Eulerian granular phase simulation**
  In the section, **Models**, the Eulerian multi-phase is set on 2 phases and the viscous model must be set to the selected numerical model. In **Materials** the materials and its parameters are set. In the section, **Phases**, the primary phase is set to the fluid phase (for example air), the **Interaction** has to be set to the chosen drag law or to the given drag law in the UDF. After that, the secondary (granular) phase must be assigned. The diameter, granular viscosity, granular conductivity, solids pressure and radial distribution function are set in the UDF methods. The packing limit for the granular medium should be set to the same value as chosen in **variables.h**. The calculation of the **Granular Temperature** should be set to "partial differential equation". Next, the **Cell Zone Conditions** are defined. For the Eulerian-Eulerian granular phase, the source terms must be activated. One source term for each of the x-Momentum, y-Momentum, z-Momentum and granular temperature is needed. For each source term, an UDF method can be found. Since a custom BC is used, the standard BC in FLUENT (Johnson and Jackson) must be deactivated. Therefore, at all walls, the specified shear and specularity coefficient must be set to zero.

- **Setup of Lagrangian simulation**
  In **Models:Discrete Phase**, no interaction or particle treatment has to be set. The necessary physical models can be set, but no collision modelling should be activated. The collision modelling is included in the hybrid model UDF. The UDF for **Body force**, **Scalar Update** and **DPM Time step** should be included. The number of scalars has to be

set to 20. After that, the injection of particles must be done. The mass flow rate should be set to the same value as set in variables.h.

- **Setup of the coupling**
  The coupling forces are computed after each time step and are set in the menu bar point **Define:User-Defined:Function Hooks:Execute at End**. Here the UDF (executeEnd) must be included. Also, a UDF to create a threshold for the granular temperature can be set here. This should only be used at the beginning of the simulation, until the granular temperature is in a stable condition.

- **Final steps before simulations start**
  The **Solution Controls** can be set as usual. It is recommended to start the simulation with a first order discretization scheme, and once a stable simulation is reached, it can be changed to either second order or QUICK scheme. Under the menu point **Define:User-Defined**, the amount of user defined memory should be set to 250. To check the simulation progress, monitors should be used. The time step size has to be defined under **Run calculation**. Now the simulation should be saved. After that the simulation can be initialized. Then with **Define:User-Defined:Execute on Demand** the UDF "INIT_UDMI" should be run. This initializes all variables in FLUENT (user defined memories) for use in the UDF. Save now again. If the simulation was set up correctly, the scheme file can be loaded and the simulation can be started with the command: "(calculateHybrid)" or if agglomeration is included "(calculateHybridAgglo)".

2. **Simulate gas flow with Eulerian granular phase for n-Timesteps**
   During this period in the simulation, the Eulerian granular phase simulation is computed for n-time steps. Computed exchange forces from older Lagrangian tracer trajectories are already included as additional source terms. At the end of each time step, the "Execute at End" UDF is run and adapts the source terms for the granular BC and for the Lagrangian tracer particles. After n time steps, the simulation is stopped and the next step is done.

3. **Calculate Tracer-particles**
   In this step, the Lagrangian tracer trajectories are computed on a fixed velocity field for both the fluid and granular phases. The exchange terms for the

Magnus force and the BC force are computed from tracer particles. Also, the mass flow through the inlet and outlet can be computed and monitored for the Lagrangian phase. This allows for the efficiency calculation of a cyclone.

4. **Update Exchange terms**

   With the newly computed forces, new source terms for the Eulerian granular phase are computed based on the given results from the tracer particles. These exchange terms are also computed after each time step, so this step is not necessary. Since computational time is expensive, and should be used sparingly, the update of the exchange terms should be computed after the tracer particle computation. Therefore it can be excluded from the adaption UDF run, which happens after each Eulerian granular phase time step.

5. $t < t_{end}$?

   If all loops in the "calc.scm" file are finished, the simulation is stopped.

6. **Final Solution**

   Now the final solution is reached and the result can be validated by the user. It is also possible to run tracer trajectories on the final result to display the particle tracks.

D

Cyclone dimensions based on Muschelknautz theory

Most of the cyclones found in industry and scientific experiment facilities are modeled with the Muschelknautz theory. A description of the modelling can be found in Muschelknautz et al. [1994] and Hoffmann and Stein [2008] (chapter 6, The Muschelknautz Method of Modelling). Also, in the work of Kahrimanovic [2009], a chapter is dedicated to cyclone design with the Muschelknautz method.

Two important standard values for a cyclone are the pressure loss and the cut-point diameter. The cut-point diameter denotes the diameter of particles which are separated with a 50% chance. The theory of Muschelknautz can be used to estimate, in an analytic way, the pressure drop and cut-point diameter of a cyclone.

Calculation of cut-point diameter

First, the necessary parameters for a cyclone must be defined. In Figure D.1, the important dimensions are illustrated. In the figure, H is the total height of the

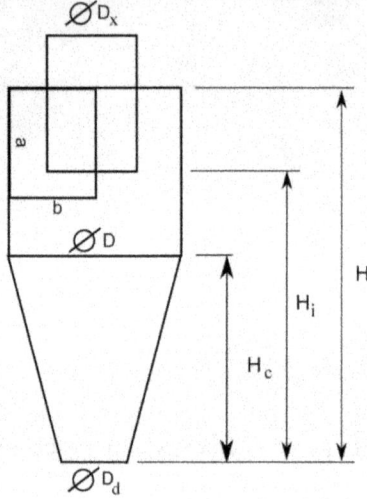

Figure D.1: Cyclone geometry with various parameters used for design [Hoffmann and Stein, 2008]

cyclone, H_c the height of the cone, with H_i the distance of the tube inside the cyclone. The inlet area is described by a and b, while D, D_x and D_d are the important diameters. The modelling of Muschelknautz starts with the calculation of an entrance "constriction coefficient", α [Hoffmann and Stein, 2008]. It is calculated with the following empirical formula

$$\alpha = \frac{1}{\xi} \left\{ 1 - \sqrt{1 + 4 \left[\left(\frac{\xi}{2}\right)^2 - \frac{\xi}{2} \right] \sqrt{1 - \frac{(1 - \xi^2)(2\xi - \xi^2)}{1 + L}}} \right\} \quad \text{(D.1)}$$

where L denotes the incoming mass loading, and

$$\xi = \frac{2b}{D} \quad \text{(D.2)}$$

denotes the ratio of the inlet width to the cyclone diameter. With the knowledge

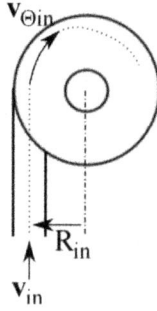

Figure D.2: Plan view of cyclone [Hoffmann and Stein, 2008]

of α, the wall velocity can be computed. Additional variables are used which are depicted in figure D.2. The wall velocity in tangential direction is given by

$$v_{\Theta,w} = \frac{2v_{in}R_{in}}{\alpha D} \tag{D.3}$$

where the mean inlet velocity. It is given by

$$v_{in} = \frac{Q}{ab} \tag{D.4}$$

where Q is the mass flow out of the vortex finder. The wall axial velocity is given by

$$v_{zw} = 0.9 \frac{Q}{\pi \left[(0.5D)^2 - (0.25D_x D)^2 \right]} \tag{D.5}$$

The factor 0.9 is necessary, because it was found by Muschelknautz and Trefz that approximately 10% of the incoming gas 'short circuits' the cyclone and flows radially inward on the top of the cyclone, down outside the vortex tube and finally in the vortex tube [Hoffmann and Stein, 2008]. Pirker et al. [2012] demonstrated this effect in simulations with a hybrid model that combines Lattice-Boltzmann modelling with standard CFD models. Now it is possible to calculate a cyclone body Reynolds number, which is defined as

$$\mathrm{Re}_R = \frac{0.25R_{in}D_x D v_{zw}\rho_g}{H\mu_g \left(1 + \left(\frac{v_{zw}}{v_{\Theta,m}} \right)^2 \right)} \tag{D.6}$$

where the term $v_{\Theta,m}$ depends on the inner vortex tangential velocity. The term $\frac{v_{zw}}{v_{\Theta,m}}$ is mostly $\ll 1$ so that it can be neglected [Hoffmann and Stein, 2008] and the cyclone body Reynolds number reduces to

$$\text{Re}_R = \frac{0.25 R_{in} D_x D v_{zw} \rho_g}{H \mu_g} \tag{D.7}$$

which can be used as a first approximation.

An important part of modelling cyclones is the wall friction factor λ. It is used to translate the impact of a rough wall into the separation efficiency and pressure loss of the cyclone. Muschelknautz et al. [1994] introduced a diagram in which the wall friction factors for cylindrical and conical cyclones, based on air, λ_{air}, can be found the Reynolds number, Re_R, and the relative wall roughness, $\gamma_{rel,w}$. Therefore the Reynolds number and the wall friction factor must be independent, which is the case if $\text{Re}_R > 2000$ [Hoffmann and Stein, 2008]. The relative wall roughness is calculated with

$$\gamma_{rel,w} = \frac{2k_s}{D} \tag{D.8}$$

where k_s denotes the absolute roughness of the inner surface of the cyclone. Hoffmann and Stein [2008] presents another possibility to get the wall friction factor, λ_{air}, based on a factor for smooth walls and an added contribution due to wall roughness. After λ_{air} is known, the total friction factor can be calculated with

$$\lambda = \lambda_{air} + 0.25 \left(\frac{D}{D_x}\right)^{-0.625} \sqrt{\frac{\eta L \text{Fr}_x \rho_g}{\rho_{str}}} \tag{D.9}$$

where η is the cyclone's overall collection efficiency. Fr_x denotes the Froude number is given by

$$\text{Fr}_x = \frac{v_x}{\sqrt{D_x g}} \tag{D.10}$$

and v_x is the superficial velocity through the inlet section of the vortex tube [Hoffmann and Stein, 2008]. ρ_{str} represents the bulk density inside the particle strand, and is given by

$$\rho_{str} = 0.3 \ldots 0.5 \alpha_{s,max} \rho_s. \tag{D.11}$$

In the next step, the velocity of the gas at the inner core, using the vortex finder tube diameter, is estimated with the following equation

$$v_{\Theta,CS} = v_{\Theta,w} \frac{\frac{D}{D_x}}{1 + \frac{\lambda A_R v_{\Theta,w} \sqrt{D/D_x}}{2Q}}$$ (D.12)

where A_R denotes the total inside area of the cyclone. This area is given by

$$A_R = A_{top} + A_{cyl} + A_{cone} + A_{vortexTube}$$

$$= \pi \left[\left(\frac{D}{2} \right)^2 - \left(\frac{D_x}{2} \right)^2 + D \left(H - H_c \right) + \right.$$

$$\left. \left(\frac{D}{2} + \frac{D_d}{2} \right) \sqrt{H_c^2 + \left(\frac{D}{2} - \frac{D_d}{2} \right)} + D_x \left(H - H_i \right) \right].$$ (D.13)

Now it is possible to calculate the cut-point diameter. The equation suggested by Hoffmann and Stein [2008] is a variation of the famous cut-point formula given by Barth [1956], and follows

$$d_{50} = d_{fact} \sqrt{\frac{18\mu_g \left(0.9Q \right)}{2\pi \left(\rho_s - \rho_g \right) v_{\Theta,CS}^2 H_i}}$$ (D.14)

where d_{fact} denotes a correction factor that may be applied to match the observed cut-point diameter in practice. This value is typically in a range of about 0.9 to 1.4 [Hoffmann and Stein, 2008]. Additionally it should be mentioned that this equation can only be used if the Stokes-law regime is present. This can be checked by computing the particle Reynolds number inside the cyclone (Hoffmann and Stein [2008]). If $Re_p < 0.5$, Stokes-law applies. This is usually the case. For other regimes, take a look at Hoffmann and Stein [2008].

Computation of pressure drop

The description of the pressure drop is given by Hoffmann and Stein [2008]. The full pressure drop in the cyclone is given by

$$\delta p = \delta p_{body} + \delta p_x + \delta p_{acc}$$ (D.15)

where

$$\delta p_{\text{body}} = \frac{\lambda A_R \rho_g \left(v_{\Theta,w} v_{\Theta,CS}\right)^{1.5}}{1.8Q} \tag{D.16}$$

defines the loss in the cyclone body.

$$\delta p_x = \left[2 + \left(\frac{v_{\Theta,CS}}{v_x}\right)^2 + 3 \left(\frac{v_{\Theta,CS}}{v_x}\right)^{\frac{4}{3}}\right] \frac{\rho_g v_x^2}{2} \tag{D.17}$$

defines the loss in the core and in the vortex finder. If the gas-solid mixture must be accelerated from a low-velocity (v_0) region to the entrance velocity (v_{in}) of the cyclone an acceleration term appears in the equation

$$\delta p_{\text{acc}} = (1 + L) \frac{\rho_g \left(v_{\text{in}}^2 - v_0^2\right)}{2}. \tag{D.18}$$

In this equation, no slip between the solid and gas phases is accounted for.

E

Nomenclature

Parameter	Name	Unit
α_g	volume fraction of gas phase	1
α_s	volume fraction of solid phase	1
$\alpha_{s,min}$	volume fraction when frictional shear appears	1
$\alpha_{s,max}$	maximum volume fraction	1
β	virtual angle at wall	rad
β	interphase momentum exchange	1/s
β	shear rate in Saffman force computation	1/s
β_p	interphase momentum exchange at particle	1/s
$\beta_{p,coll}$	collisional interphase momentum exchange	1/s
β_{poly}	poly-disperse interphase momentum exchange	1/s
β_s	interphase momentum exchange at solid phase	1/s
β_0	tangential coefficient of restitution	1/s
ϵ	dissipation	m^2/s
$\eta_{i,j}$	collisions probability	1
γ_Θ	Dissipation of PTE by collisional viscous damping	kg/ms^2

123

μ	wall coefficient of friction	kg/(ms)
μ_w	wall coefficient of friction	kg/(ms)
μ_i	coefficient of internal friction	1
μ_s	solids shear viscosity	kg/(ms)
$\mu_\mathrm{s}^\mathrm{kc}$	solids kinematic and collisional shear viscosity	kg/(ms)
$\mu_\mathrm{s,coll}$	solids collisional shear viscosity	kg/(ms)
$\mu_\mathrm{s,kin}$	solids kinematic shear viscosity	kg/(ms)
$\mu_\mathrm{s}^\mathrm{fr}$	solids frictional shear viscosity	kg/(ms)
μ_w	wall coefficient of friction	kg/(ms)
λ	mean free path in Knudsen number	m
$\lambda_\mathrm{s}^\mathrm{kc}$	solids bulk viscosity	kg/(ms)
ρ_g	density of gas phase	kg/m^3
ρ_p	density of particle	kg/m^3
ρ_s	density of solid phase	kg/m^3
ϕ_i	angle of internal friction	rad
ϕ_s	PTE production between solid and gas phase	kg/ms^2
ϕ'	specularity coefficient	1
Θ	granular temperature	m^2/s^2
τ	shear rate	kg/(ms^2)
τ_k	relaxation time	s
$\tau_\mathrm{s}^\mathrm{kc}$	kinetic shear stress (to wall)	kg/(ms^2)
$\tau_\mathrm{s}^\mathrm{fr}$	frictional shear stress	kg/(ms^2)
ω	absolute value of particle angular velocity	1/s
ω_p	angular velocity of particle	1/s
$\mathbf{\Omega}_\mathrm{p}$	particle fluid rotation velocity	1/s
$\mathbf{A}$	transformation matrix	1
a_c	affected boundary area	m^2
C	constant	1
C_D	drag coefficient	1
C_L	rotation coefficient	1
C_LS	dimensionless description of Saffman force	1
C_R	drag coefficient of rotation	1
c_i	volume fraction of particle class i	1
D	dimension	1
D_Θ	dissipation of pseudo thermal temperature	1
D_g	rate-of-strain tensor for gas phase	1/s
D_s	rate-of-strain tensor for solid phase	1/s
d_p	particle diameter	m

$d_{\mathrm{p,max}}$	maximum assumed particle diameter	m
d_s	solids particle diameter	m
d_{strand}	particle strand diameter	m
E_{after}	energy after particle collision	$\mathrm{kg\ m^2/s^2}$
E_{before}	energy before particle collision	$\mathrm{kg\ m^2/s^2}$
E_{mech}	energy dissipation by deformation	$\mathrm{kg\ m^2/s^2}$
E_{vW}	energy dissipation by van-Waals-force	$\mathrm{kg\ m^2/s^2}$
e	restitution coefficient	1
e_{n}	restitution coefficient normal to wall	1
e_{w}	particle-wall restitution coefficient	1
e_{p}	particle-particle restitution coefficient	1
e_s	solids restitution coefficient	1
$e_{\mathrm{s,p}}$	solids-particle restitution coefficient	1
e_{t}	restitution coefficient tangential to wall	1
$\mathbf{F}_{\mathrm{Basset}}$	Basset force	$\mathrm{kg\ m/s^2}$
$\mathbf{F}_{\mathrm{p}}$	particle force	$\mathrm{kg\ m/s^2}$
$\mathbf{F}_{\mathrm{p}}$	pressure gradient force	$\mathrm{kg\ m/s^2}$
$\mathbf{F}_{\mathrm{Saff}}$	Saffman force	$\mathrm{kg\ m/s^2}$
$\mathbf{F}_{\mathrm{VM}}$	virtual mass force	$\mathrm{kg\ m/s^2}$
f_{coll}	collisions rate, frequency	$1/\mathrm{s}$
$\mathbf{f}_{\mathrm{g,add}}$	additional forces in gas phase	$\mathrm{m/s^2}$
$\mathbf{f}_{\mathrm{p}}$	force per mass	$\mathrm{m/s^2}$
$\mathbf{f}_{\mathrm{p,drag}}$	drag force per mass	$\mathrm{m/s^2}$
$\mathbf{f}_{\mathrm{p,magnus}}$	Magnus force per mass	$\mathrm{m/s^2}$
$\mathbf{f}_{\mathrm{p,press}}$	granular pressure force on particle	$\mathrm{m/s^2}$
$\mathbf{f}_s$	force per mass, Eulerian granular phase	$\mathrm{kg/ms^2}$
$\mathbf{f}_{\mathrm{s,add}}$	additional forces in solid phase	$\mathrm{kg/ms^2}$
$\mathbf{f}_{\mathrm{s,wall}}$	solid-particle exchange force at wall	$\mathrm{kg/ms^2}$
$G_{\mathrm{i,l,m}}$	volume creation rate in i by agglomeration of l and m	$\mathrm{m^3/s}$
$\mathbf{g}$	gravitation	$\mathrm{m/s^2}$
g_0	radial distribution function	1
$H_{i,j}$	collisions probability	1
h_i	amount of particles with diameter d_i	1
$h_{i,\mathrm{agg}}$	amount of particles d_i after agglomeration	1
$\mathbf{I}$	unity matrix	1
I	inertia	$\mathrm{kgm^2}$
I_{p}	inertia of particle	$\mathrm{kgm^2}$
l	distance	m

k	turbulent kinetic energy	$\mathrm{m^2/s^2}$
k_B	Boltzmann constant	$\mathrm{J/K}$
$K_{i,j}$	agglomeration rate between class i and j	$\mathrm{m^3/s}$
Kn	Knudsen number	1
K_strand	ratio between max. particle and strand diameter	1
K_poly	poly-disperse drag cofficient	1
l_s	particle mean free path	m
L_C	characteristic length scale	m
M_x	momentum in x-direction	$\mathrm{kg\,m/s}$
M_y	momentum in y-direction	$\mathrm{kg\,m/s}$
m_p	particle mass	kg
m	mass	kg
N	number of particles	1
N	normal stress	$\mathrm{kg/ms^2}$
$\mathbf{n}$	normal vector to wall	1
$\dot{n}_\mathrm{s,p}$	particle collision frequency in volume	$\mathrm{m^3/s}$
p	pressure of gas phase	Pa
p_s	granular pressure	$\mathrm{kg/ms^2}$
p_s^kc	kinetic and collisional granular pressure	$\mathrm{kg/ms^2}$
q	flux between particle and wall	$\mathrm{kg/m^2s}$
$\mathbf{q}$	pseudo thermal energy	$\mathrm{m/s^2}$
$\mathbf{n}\cdot\mathbf{q}$	flux of fluctuation energy at boundary wall	$\mathrm{kg/s^3}$
r	radius of particle	m
r	magnitude of mean velocity in boundary condition	$\mathrm{m/s}$
Re	Reynolds number	1
Re_β	Reynolds number for deviation of velocities	1
Re_p	particle Reynolds number	1
Re_R	rotational particle Reynolds number	1
s	particle surface to surface distance	m
$\mathbf{S}_\mathrm{s}^\mathrm{kc}$	kinetic solids stress tensor	$\mathrm{1/s^2}$
$\mathbf{S}_\mathrm{s}^\mathrm{fr}$	frictional solids stress tensor	$\mathrm{1/s^2}$
S	shear stress	$\mathrm{kg/ms^2}$
$\mathrm{St}_\mathrm{p,k}$	Stokes number	1
T	temperature	Kelvin or Celsius
t	time	s
t_0	time when particle enters inlet	s
t_agg	time needed for agglomeration	s
$\mathbf{t}$	angular momentum, torque	$\mathrm{1/s^2}$

$\mathbf{t}_{p,g}$	particle torque by gas phase	$1/s^2$
$\mathbf{t}_{p,add}$	additional particle torque	$1/s^2$
$\mathbf{T}_g$	gas stress tensor	$1/s^2$
$\langle u \rangle$	average absolute velocity	m/s
u	absolute velocity	m/s
u_{crit}	critical velocity for agglomeration	m/s
$\bar{u}_s^{sl}$	average slip velocity	m/s
$\mathbf{u}_p$	particle velocity	m/s
$\bar{\mathbf{u}}_p$	average particle velocity	m/s
$\mathbf{u}_p'$	stochastic fluctuation of particle velocity	m/s
$\mathbf{u}_{p,rebound}$	rebound particle velocity after wall collision	m/s
$\mathbf{u}_g$	velocity of gas phase	m/s
$\mathbf{u}_{n,after}$	normal particle velocity normal to wall after collision	m/s
$\mathbf{u}_{n,before}$	normal particle velocity normal to wall before collision	m/s
$\mathbf{u}_s$	velocity of gas phase	m/s
$\mathbf{u}_s^{sl}$	slip velocity, between particle and wall	m/s
$\mathbf{u}_{s,rebound}$	rebound solid velocity after wall collision	m/s
$u_{s,nW}$	fluctuating rebound velocity after wall collision	m/s
$\mathbf{u}_{t,after}$	tangential particle velocity after wall collision	m/s
$\mathbf{u}_{t,before}$	tangential particle velocity before wall collision	m/s
V	volume	m^3
V_{grid}	cell volume	m^3
w_a	relative particle velocity from acceleration	m/s
x	average distance between particles and wall	m

Table E.1: Nomenclature

List of Figures

1.1 Picture of a preheater tower 2

1.2 Sketch of particle-laden flow regimes 5

1.3 Sketch of single particle that is followed by time. 6

1.4 Sketch of granular flow through fixed control volume 7

1.5 Sketch of particle-laden flow regimes; depicting the applicability of (I) Lagrangian particle tracing, (II) Eulerian particulate phase model and (III) hybrid particle-laden flow modelling 8

1.6 Organization of this book 10

2.1 Range of application of Eulerian-Eulerian granular phase model. . . . 12

2.2 Schematic of Eulerian granular phase model, without considering any boundary conditions. The grayed ellipse describes the gas or granular phase. Inside each ellipse the computed variables are shown. Each phase model is described by a set of balancing equations which are represented by the grayed rectangles. Further models, represented by white ellipses, provide information for the balancing equations. 13

2.3 Simple model of particle packing [Sinclair and Jackson, 1989, Gidaspow and Huilin, 1998] 17

2.4 Comparison of the different models for the radial distribution function and measurements 19

2.5 Schneiderbauer et al. [2012b] boundary conditions: Stress ratio $R = \frac{7}{2}\frac{1+e_w}{1+\beta_0}\frac{S}{N}$ for different tangential restitution coefficients β_0 over normalized slip $r = \frac{u_{sl}}{\sqrt{3\Theta}}$. The oblique and horizontal dotted lines represent Jenkins no sliding and all sliding limits. The $\circ$, $\triangle$, $\square$ and $\Diamond$ symbols indicate results from simulations done by Louge [1994]. 29

2.6   Normalized flux of fluctuation energy, $q/\left(\sqrt{3\Theta}N\right)$ of the three models, Johnson and Jackson [1987], Jenkins and Louge [1997] and Schneiderbauer et al. [2012b] over normalized slip $r = \frac{u_{sl}}{\sqrt{3\Theta}}$.   30

3.1   Range of validity of the discrete phase model   33
3.2   Forces and parameters of a discrete phase particle   34
3.3   Drag force on the particle   34
3.4   Magnus force   36
3.5   Saffman force   36
3.6   Example for different restitution coefficients   42
3.7   Particle wall collision with rough wall. The difference of the normal vector between flat wall n_c and the rough wall n_l is shown.   42
3.8   Virtual wall inclination angle β. The particle can hit the front or backside of a wall triangle element.   43
3.9   Shadow effect in the wall collision model   43
3.10   If the shadow effect is considered, the probability function for the inclination angle tends to values for front side collision. In this Figure an example is shown, where a back side collision is not possible. . . .   44

4.1   Overview of EUgran+Poly model. On the right side is the Eulerian granular phase model with the considered important physical effects depicted. On the left side is the Eulerian-Lagrangian discrete phase model.   47
4.2   The Lagrangian tracer particle cloud gives a detailed look into the granular phase.   49
4.3   Homogeneous and heterogeneous particle distribution in a CFD grid cell.   51
4.4   Particle strand with different diameters and slipstream effect on gas velocity inside the strand. $\mathbf{u}_g$ denotes the computed gas velocity in the grid cell.   51
4.5   Illustration of granular pressure.   54
4.6   The flowchart for the simulation with the hybrid model. An Eulerian granular phase simulation is computed for N time steps. After that, the tracer particles are computed with additional information gained by the values of the solid phase. After their computation, the source force fields from the Langrangian tracer are computed and smoothed. After that the Eulerian granular simulation and the whole process is redone.   55

5.1 Simple sketch of the assumed agglomeration process 57

5.2 Discretized particle diameter distribution. The arrows show in which class two agglomerating particles will create a new particle. $\otimes$ indicates that aggregates which are bigger than the threshold diameter are not considered. 62

5.3 Influence of turbulence to the particle movement by Stokes number. . 65

5.4 Schematic diagram, showing the agglomeration rate of the different models for a $d_{Ref} = 1\,\mu m$ and $St_j = [0, 25500]$. The impact of Brownian agglomeration decreases with increasing diameter, while the agglomeration based on acceleration and shear increases. 69

5.5 Schematic illustration of collision in a turbulent fluid flow. If the collision partner is inside a defined area before collision, the collision happens. Otherwise the particle flows around the collision partner, due to the effect of the surrounding fluid. 70

5.6 Bus stop model sequence: Particle cloud with diameter d_i is traced and finds two particles d_j and d_{i-j} which agglomerate to a particle d_i. These are added to the traced cloud. 73

5.7 Block diagram of the Bus stop Model. The model consists of three main steps. First, the pre-computation step for information creation. Second, the computation of agglomeration rates. Thirdly, the simulation of agglomeration with post-correction of mass. 75

5.8 Simple sketch of the assumed agglomeration process by the volume population balance model 76

6.1 Geometry of duct bend [Kuan et al., 2007] 80

6.2 Comparison of the tangential velocities of the gas phase for the hybrid model and measurements at three different positions in the bend. . . 82

6.3 Comparisons of the tangential velocities for the mono-dispersed Eulerian granular model and the poly-dispersed EUgran+Poly model at three different in the bend. 83

6.4 Particle size distribution in the bend. At the surface of the particle strand a bigger diameter as in the particle strand at the wall can be seen. 84

6.5 The wind tunnel in our academic laboratory. Most of our experiments are done at this facility. 85

6.6 Picture of the double-loop geometry. The double-loop geometry was installed in the wind tunnel and used for several experiments in our laboratory. 86

6.7 Double-loop geometry with velocity measurement positions [Pirker
     et al., 2010]. 87
6.8 Comparison of experiment and simulation. Both show the dispersion
     of the particle strand behind the second loop. 88
6.9 Average particle velocity at position 1, 2 and 3 after the second loop. 88
6.10 Comparison of measured and simulated particle parameters. Volume
     fractions were compared at two positions behind the second loop. . . 89
6.11 Sketch of the vertical pipe, which was used to validate the agglomer-
     ation model 91
6.12 Comparison of particle volume fraction between discretized Smolu-
     chowski model and the Bus stop model (in MATLAB and FLUENT) 92
6.13 Change of particle size distribution with height ($\bigcirc$ 1 m,$\square$ 5 m, $\lozenge$ 25 m
     and $\triangleright$ 50 m) in the vertical pipe. Comparison between Smoluchowski
     and Bus stop model. 92

7.1 Geometry of the in-house cyclone. 94
7.2 Picture of the in-house cyclone. 95
7.3 Geometry of the limestone separation cyclone. 96
7.4 Separation efficiency compared to Muschelknautz theory for a mass
     loading $L = 0.01$ 98
7.5 Volume fraction in the cyclone. The volume fraction is scaled loga-
     rithmically. 98
7.6 Volume fraction, α_s, in wall adjacent cells for $L = 1$. The particle
     strand formation can be seen. 99
7.7 Comparison of separation efficiency for different numerical models for
     mass loading $L = 1$ 100
7.8 Volume fraction of granular material in the cyclone. On the left side,
     the Eulerian-Eulerian granular phase and on the right side the coupled
     Lagrangian tracer trajectories can be seen. 101
7.9 Difference of particle size distribution at the inlet and at the particle
     outlet of the cyclone. 102

C.1 Connection between files and FLUENT. 113

D.1 Cyclone geometry with various parameters used for design [Hoffmann
     and Stein, 2008] 118
D.2 Plan view of cyclone [Hoffmann and Stein, 2008] 119

List of Tables

2.1   Radial distribution functions 18

6.1   Simulation parameters used in the duct bend simulation. 79

6.2   Particle diameter distribution [Kuan et al., 2007] 81

6.3   Simulation parameters for double-loop geometry 87

7.1   Simulation parameters for the in-house cyclone 95

7.2   Particle diameter distribution for glass particle 96

7.3   Simulation parameters for the limestone separation cyclone 97

7.4   Particle diameter distribution for limestone particles 97

7.5   Comparison of computational effort, averaged over 1000 time steps . . 101

E.1   Nomenclature 127

Bibliography

M. Abedi. Effect of restitution coefficient on inertial particle separator's efficiency. Mechanical engineering master's theses, IRis Northeastern University, 2009.

J. Abrahamson. Collision rates of small particles in a vigorously turbulent fluid. *Chemical Engineering Science*, 30:1371–1379, 1975.

J. Abrahamson, R. Jones, A. Lau, and S. Reveley. Influence of entry duct bends on the performance of return-flow cyclone dust collectors. *Powder Technology*, 123: 126–137, 2002.

K. Agrawal, N. Loezos, M. Syamlal, and S. Sundaresan. The role of meso-scale structures in rapid gas-solid flows. *Journal of Fluid Mechanics*, 445:151–185, October 2001.

B. J. Alder and T. E. Wainwright. Studies in Molecular Dynamics: II. Behaviour of a Small Number of Elastic Spheres. *J. Chem. Phys.*, 33:1469, 1960.

H. C. Anh. *Modellierung der Partikelagglomeration im Rahmen des Euler/Lagrange-Verfahrens und Anwendung zur Berechnung der Staubabscheidung im Zyklon*. Phd, Haale, 2004.

Inc. ANSYS. *ANSYS FLUENT 12.0 - Theory Guide*. ANSYS, April 2009.

A. Bakker. Applied computational fluid dynamics. Lecture 14 - Multiphase Flows, 2008.

H. M. Barkla. The magnus or robins effect on rotating spheres. *Journal aof Fluid Mechanics*, 47:437-447, 1971.

W. Barth. Brennstoff-Wärme-Kraft (in German). *Heft 1*, 8, 1956.

A. B. Basset. Treatise on hydrodynamics. *Deighton Bell, London*, 2:285-297, 1881.

J. Boussinesq. Theorie analytique de la chaleur. *L'École Polytechnique, Paris*, 2: 224, 1903.

Brilliantov, N. V. and Pöschel, T. *Kinetic Theory of Granular Gases*. Oxford Graduate Texts, 2004.

N. F. Carnahan and K. E. Starling. Equations of State for Non-Attracting Rigid Spheres. *J. Chem. Phys.*, 51:635, 1969.

S. Chialvo, J. Sun, and S. Sundaresan. Bridging the rheology of granular flows in three regimes. *Physical Review*, E 85:021305, 2012.

J. H. Conway and N. J. A. Sloane. *Sphere Packings, Lattices and Groups*. Springer Verlag, New York, 2nd edition, 1993.

C. Crowe, M. Sommerfeld, and Y. Tsuji. *Multiphase Flows with Droplets and Particles*. CRC Press, 1998.

C. T. Crowe, M. P. Sharma, and D. E. Stock. The Particle-Source-In Cell (PSI-CELL) Model for Gas-Droplet Flows. *Journal of Fluids Engineering*, June 1977.

P.A. Cundall and M.P. Strack. A discrete numerical model for granular assemblies. *Geotechnique*, 29:47-65, 1979.

E. Cunningham. On the Velocity of Steady Fall of Spherical Particles through Fluid Medium. *Proc. Roy. Soc.*, A 83, 1910.

F. da Cruz, S. Emam, M. Prochnow, J.-N. Roux, and Chevior F. Rheo-physics of dense granular materials: Discrete simulation of plane shear flows. *Physical Review*, E 72:021309, 2005.

D. S. Dandy and H. A. Dwyer. A sphere in shear flow at finite Reynolds number: effect of shear on a prticle lift, drag, and heat transfer. *Journal of Fluid Mechanics*, 216:381–410, 1990.

S. Dartevelle. *Numerical and Granulometric Approaches to Geophysical Granular Flows*. PhD thesis, Michigan Technological University, 2003.

F C. de Almeida. The collisional problem of cloud droplets moving in a turbulent environment. part ii: Turbulent collision efficiencies. *Journal of atmospheric sciences*, 36:1564–1576, 1979.

S.C.R. Dennis, S.N. Singh, and D.B. Ingham. The steady flow due to a rotating sphere at low and moderate Reynolds numbers. *Journal of Fluid Mechanics*, 101: 257–279, 1980.

J. Ding and D. Gidaspow. A Bubbling Fluidization Model Using Kinetic Theory of Granular Flow. *AIChE J.*, 36(4):523–538, 1990.

S. Elghobashi. On predicting particle-laden turbulent flows. *Applied Scientific Research*, 52:309–329, 1994.

S. Ergun. Fluid Flow through Packed Columns. *Chem. Eng. Prog.*, 48 (2):89–94. 1952.

FLUENT. *Fluent 6.3 User's Guide*. Fluent Inc., Lebanon, 2006.

J. Forterre and O. Pouliquen. Flow of Dense Granular Media. *Annual Review of Fluid Mechanics*, 40:1–24, 2008.

Th. Frank. *Parallele Algorithmen für die numerische Simulation dreidimensionaler, disperser Mehrphasenströmungen und deren Anwendungen in der Verfahrenstechnik*. PhD thesis, Aachen, 2002. Habilitation thesis.

D. Gidaspow. *Multiphase Flow and Fluidization - Continuum and Kinetic Theory Descriptions*. Academic Press, SanDiego, 1994.

D. Gidaspow and L. Huilin. Equation of State and Radial Distribution Functions of FCC Particles in a CFB. *AIChE J.*, 44:279, 1998.

D. Gidaspow, R. Bezburuah, and J. Ding. Hydrodynamics of Circulating Fluidized Beds, Kinetic Theory Approach. In *Fluidization VII, Proceedings of the 7th Engineering Foundation Conference on Fluidization*, pages 75–82, 1992.

I. Goldhirsch. Introduction to granular temperature. *Powder Technology*, 182:130–136, 2008.

C. Goniva, C. Kloss, N.G. Deen, J.A.M. Kuipers, and S. Pirker. Influence of Rolling Friction Modelling on Single Spout Fluidized Bed Simulations. *Particuology*, 2012. doi: 10.1016/j.partic.2012.05.002.

A. D. Gosman and E. Ioannides. Aspects of computer simulation of liquid-fuelled combustors. *J. Energy*, 7(6):482–490, 1983.

R. Hiller and F. Löffler. Der Einfluß von Partikelstoß und Partikel-Haftung auf die Abscheidung von Partikeln in Faserfiltern. *Chemie Ingeneur Technik*, 52:352–353, 1980.

J. O. Hinze. *Turbulence*. McGraw Hill, New York, 2 edition, 1975.

A. C. Hoffmann and L. E Stein. *Gas cyclones and swirl tubes*. Springer, 2nd edition, 2008. ISBN 978-3-540-74694-2.

M. J. Hounslow, R. L. Ryall, and V. R. Marshall. A discretized population balance for nucleation, growth and aggregation. *AiChe*, 34, 1988.

C. M. Hrenya and J. L. Sinclair. Effects of particle-phase turbulence in gas-solid flows. *AIChE Journal*, 43(4):853–869, April 1997.

K. Hui, P. K. Haff, J. E. Ungar, and R. Jackson. Boundary Conditions for high-shear grain flows. *Journal of Fluid Mechanics*, 145:223–233, 1984.

L. Huilin, D. Gidaspow, J Bouillard, and L. Wentie. Hydrodynamic Simulation of Gas-Solid Flow in a Riser Using Kinetic Theory of Granular Flow. *Chem. Eng. J.*, 95:1–13, 2003.

B. Hussmann, M. Pfitzner, Th. Esch, and Th. Frank. A stochastic particle-particle collision model for dense gas-particle flows implemented in the Lagrangian solver of ANSYS CFX and its validation. In *6th International Conference on Multiphase Flows*, Leipzig, Germany, 2007. ICMF, ICMF.

H. Iddir and H. Arastoopour. Modeling of multitype particle flow using the kinetic theory approach. *AIChE J.*, 6:1620–1632. 2005.

M. Ishii. Thermo-fluid dynamic theory of two-phase flow. *Collection de la Direction des Etudes et recherches d'Electricite de France*, 1975.

R. Jackson. The dynamics of Fluidized Particles. *Cambridge Monographs on Mechanics*, page 210. 2000.

J. T. Jenkins. Boundary conditions for rapid granular flows: Flat, frictional walls. *Journal of Applied Mechanics*, 59:120–127, 1992.

J. T. Jenkins and M. Y. Louge. On the flux of fluctuation energy in a collisional grain flow at a flat, frictional wall. *Physics of Fluids*, 9(10):2835–2840, 1997.

James Jenkins. Boundary Conditions for Collisional Grain Flows at Bumpy, Frictional Walls. In Thorsten Pöschel and Stefan Luding, editors, *Granular Gases*, volume 564 of *Lecture Notes in Physics*, pages 125–139. Springer Berlin / Heidelberg, 2001. ISBN 978-3-540-41458-2.

P. C. Johnson and R. Jackson. Frictional-collisional constitutive relations for granular materials, with application to plane shearing. *Journal of Fluid Mechanics*, 176:67–93. 1987.

P. C. Johnson, P. Nott, and R. Jackson. Frictional-collisional equations of motion for particulate flows and their application to chutes. *Journal of Fluid Mechanics*, 210:501–535, 1990.

P. Jop, Y. Forterre, and O. Pouliquen. A consitutive law for dense granular flows. *Nature*, 441:727–730, 2006.

D. Kahrimanovic. *Numerische Simulation und experimentelle Validierung von Gas-Partikel-Strö$\frac{1}{2}$mungen*. Phd, Johannes Kepler University, Altenbergerstreet 69. 2009.

V. Kerminen. Simulation of Brownian Coagulation in the Presence of van der Waals Forces and Viscous Interactions. *Aerosol Science and Technology*, 20:2:207–214, 1994.

C. Kloss. *LIGGGHTS - A New Open Source DEM Code Applied to the Corex Process*. Phd thesis, Johannes Kepler University, 2011.

D.L. Koch. Kinetic theory for a monodisperse gas-solid suspension. *Phys. Fluids*, A 2:1723 – 1723, 1990.

W. Krainz. Entwurf und Aufbau eines Strömungskanals für granulare Strömungen. diploma thesis, Johannes Kepler University, Feb. 2007.

F. E. Kruis and K. A. Kuster. The Collision Rate of Particles in Turbulent Flow. *Chem. Eng. Comm.*, 158:201–230, 1997.

B. Kuan, W. Yang, and M. P. Schwarz. Dilute gas-solid two-phase flows in a curved 90° duct bend: CFD simulation with experimental validation. *Chemical Engineering Science*, 62:2068–2088, 2007.

S. Kumar and D. Ramkrishna. On the solution of population balance equations by discretization - III. Nucleation, growth and aggregation of particles*. *Chem. Eng. Sc.*, 52 (24):4659 – 4679, 1997.

T. Li and S. Benyahia. Revisiting Johnson and Jackson boundary conditions for granular flows. *AIChE Journal*, 58:2058–2068, 2012.

M. Y. Louge. Computer simulations of rapid granular flows of spheres interacting with a flat, frictional boundary. *Physics of Fluids*, 6(7):2253–2269, 1994.

C. K. K. Lun and S. B. Savage. The Effects of an Impact Velocity Dependent Coefficient of Restitution on Stresses Developed by Sheared Granular Materials. *Acta Mech.*, 63:15, 1986.

C. K. K. Lun, S. B. Savage, D. J. Jeffrey, and N. Chepurnity. Kinetic theories for granular flow: inelastic particles in Couette flow and slightly inelastic particles in a general flowfield. *Journal of Fluid Mechanics*, 140:223–256, April 1984.

G. Magnus. Ueber die Abweichung der Geschosse, und: Ueber eine auffallende Erscheinung bei rotirenden Kï¿½rpern. *Aus d. Abhandl. d. K. Akad. zu Berlin*, 1: 1 29, 1852.

J. B. McLaughlin. Inertial migration of a small spherre in linear shear flows. *Journal of Fluid Mechanics*, 224:261 274, 1991.

R. Mei. An approximate expression for the shear lift force on a spherical perticle at finite Reynolds number. *Int. Journal of Multiphase Flow*, 18:145 147, 1992.

C. J. Meyer and D. A. Deglon. Particle collision modeling - A review. *Minerals Engineering*, 24:719 730, 2011.

K. Mohanarangam, Z. F. Tian, and J. Y. Tu. Numerical simulation of turbulent gas-particle flow in a 90ï¿½ bend: Eulerian-Eulerian approach. *Computers and Chemical Engineering*, 2007.

K. Mohanarangam, H. J. Yang, W. anf Zhang, and J. Y. Tu, editors. *Effect of particles in a turbulent gas-particle flow within a 90° bend*, 2009. CSIRO Australia, CSIRO Australia.

E. Muschelknautz, W. Krammrock, and H. P. Schlag. Druckverlust bei der pneumatischen Förderung. *VDI - Wärmeatlas*, Abschnitt Lh:VDI Verlag, 1994.

U. Muschelknautz. Pressure Drop in Pneumatic Conveying Systems. *VDI Heat Atlas*, 2010. MK Engineering, Innsbruck, Austria.

I. Newton. A letter of Mr. Isaac Newton, of the University of Cambridge, containing his new theory about light and color. *Philosophical Transactions of the Royal Society*, pages 3075 3087, 1671-72.

B. Oesterlé. Simulations of particle-to-particle interactions in gas-solid flows. *Int. J. of Multiphase Flow*, 19(1):199 211, 1993.

S. H. Park, F. E. Kruis, K. W. Lee, and Fissan H. Evolution of Particle Size Distributions due to Turbulent and Brownian Coagulation. *Aerosol Science and Technology*, 36:419 432, 2002.

M. Pinsky, A. Khain, and M. Shapiro. Collisions of Small Drops in a Turbulent Flow. Part I: Collision Efficiency. Problem Formulation and Preliminary Results. *Journal of atmospheric sciences*, 56:2585–2600, 1999.

S. Pirker and D. Kahrimanovic. Numerical simulation of the inlet duct geometry influence in highly laden cyclones. *American Separation and Filtration Society*, 2007.

S. Pirker, D. Kahrimanovic, C. Kloss, B. Popoff, and M. Braun. Simulating coarse particle conveying by a set of Eulerian, Lagrangian and hybrid particle models. *Powder Technology*, 204(2-3):203–213, 2010.

S. Pirker, D. Kahrimanovic, and C. Goniva. Improving the applicability of discrete phase simulations by smoothening their exchange fields. *Applied Mathematical Modelling*, 35(5):2479–2488, May 2011.

S. Pirker, C. Goniva, C. Kloß, S. Puttinger, P. Seil, and S. Schneiderbauer. Locally resolving large scale turbulent structures by a hybrid - Lattice Boltzmann and Finite Volume - turbulence model. In *CD-ROM Proc. of the 6th European Congress on Computational Methods in Applied Sciences and Engineering*, University of Technology, Austria, Vienna, 2012. ECCOMAS.

Pöschel, T. *Dynamik granularer Gase*. Logos Verlag Berlin, 2000. ISBN 3-8325-0094-4.

K. K. Rao and P. R. Nott. *An Introdution to Granular Flow*. Cambrige University Press, 2008.

W. C. Reade and L. R. Collins. Effect of preferential concentration on turbulent collision rates. *Physics of Fluids*, 12 (10):2530–2540, 2000.

P. G. Saffman. The lift on a small sphere in a slow shear flow. *Journal aof Fluid Mechanics*, 22:385–400, 1965.

P. G. Saffman and J. S. Turner. On the collision of drops in turbulent clouds. *Journal of Fluid Mechanics*, 1:16–30, 1956.

O. Sawatzki. Das Strömungsfeld um eine rotierende Kugel. *Acta Mechanica.* 9:159–214, 1970.

D.G. Schaeffer. Instability in the evolution equations describing incompressible granular flow. *Journal of Differential Equations*, 66:19–50, 1987.

D. Schellander, D. Kahrimanovic, and S. Pirker. Studies of different numerical models for a turbulent particulate flow in a square pipe with 90 degree bend. In *Computational Methods in Multiphase Flow VI*, pages 57–68, Kos, 2011. WIT Press.

L. Schiller and Z. Naumann. *Ver. Deutsch. Ing.*, 77:318, 1935.

S. Schneiderbauer and S. Pirker. CFD study of a single-spout pseudo-2D bed: the impact of the solids wall boundary conditions. In M. K. Jha, M. Lazard, A. Zaharim, and K. Sopian, editors, *Proceedings of the 9th WSEAS International Conference on Fluid Mechanics (FLUIDS '12)*, pages 150–155, Harvard, Cambridge, MA, 2012a. WSEAS.

S. Schneiderbauer and S. Pirker. A frictional-kinetic model for dilute to dense gas-particle flows. In *Proceedings ECCOMAS 2012*, page not known now, not known now, 2012b.

S. Schneiderbauer, A. Aigner, and S. Pirker. A comprehensive frictional-kinetic model for gas-particle flows: Analysis of fluidized and moving bed regimes. *Chemical Engineering Science*, 80:279–292, 2012a.

S. Schneiderbauer, D. Schellander, A. Loederer, and S. Pirker. Non-steady state boundary conditions for collisional granular flows at flat frictional moving walls. *International Journal of Multiphase Flow*, 2012b. (online available at http://dx.doi.org/10.1016/j.ijmultiphaseflow.2012.03.006).

G. Schuch and F. Löffler. Über die Abscheidewahrscheinlichkeit von Feststoffpartikeln an Tropfen in einer Grasströmung durch Trägheitseffekte. *Verfahrenstechnik*, 12:302–306, 1975.

J.L. Sinclair and R. Jackson. Gas-Particle FLow in a Vertical Pipe with Particle-Particle Interactions. *AIChE J.*, 35:1473, 1989.

M. Sitarski and J. H. Seinfeld. Brownian coagulation in the transition regime. *J. Colloid. Interface Sci.*, 61, 1977.

M. Z. Smoluchowski. Versuch einer mathematischen Theorie der kogulationskinetic kolloid Lï¿½sungen. *Phys. Chem.*, 92:129–168, 1917.

M. Sommerfeld. *Modellierung und numerische Berechnung von partikelbeladenen Strömungen mit Hilfe des Euler-Lagrange-Verfahrens*. PhD thesis, Aachen, 1996. Habilitation thesis.

M. Sommerfeld. *Theoretical and Experimental Modelling of Particulate Flows*. Karman Institute for Fluid Dynamics, 2000.

M. Sommerfeld and N. Huber. Experimental analysis and modelling of particle-wall collisions. *International Journal of Multiphase Flow*, 25(6-7):1457 – 1489, 1999. ISSN 0301-9322. doi: 10.1016/S0301-9322(99)00047-6. URL http://www.sciencedirect.com/science/article/pii/S0301932299000476.

Martin Sommerfeld. Analyse der Wandkollision von nicht-sphärischen Feststoffpartikeln. *Chemie Ingenieur Technik*, 73, Dezember 2001.

A. Srivastava and S. Sundaresan. Analysis of a frictional-kinetic model for gas-particle flow. *Powder Technology*, 129:72–85, 2003a.

A. Srivastava and S. Sundaresan. Analysis of a frictional-kinetic model for gas-solid flow. *Powder Technology*, 129:72–85, 2003b.

R. B. Stull. *Meteorology for Scientists and Engineers*. Ed. Brooks/Cole, second edition, 2000. ISBN 0-534-37214-7.

M. Syamlal. The particle-particle drag term in a multi particle model of fluidization. *National Technical Information Service*, 1987.

Y. Tsuji, N. Y. Shen, and Y. Morikawa. Numerical Simulation of Gas-Solid Flows. I - Particle-to-Wall Collision. *Technology report of the Osaka university*, 39(1975): 233–241, October 1989.

142

M. A. van der Hoef, M. van Sint Annaland, N. G. Deen, and J. A. M. Kuipers. Numerical Simulation of Dense Gas-Solid Fluidized Beds: A Multiscale Modeling Strategy. *Annu. Rev. Fluid Mech.*, 40:40–70, 2008.

B. G. M. van Wachem, J. C. Schouten, and van den Bleek C. M. Comperative Analysis of CFD Models of Dense Gas-Solid Systems. *AIChE Journal*, (5):1035–1051, May 2001.

L.-P. Wang, O. Ayala, S. E. Kasprzak, and W. W. Grabowski. Theoretical formulation of collision rate and collision efficiency of hydrodynamically interacting cloud droplets in turbulent atmosphere. *Journal of atmospheric sciences*, 62:2433–2450. 2005.

C. Y. Wen and Y. H. Yu. Mechanics of Fluidization. *Chemical Engineering Progress Symposium Series*, 62:100, 1966.

J. J. E. Williams and R. I. Crane. Particle Collision Rate in Turbulent Flow. *International Journal of Multiphase Flow*, 9:421–435, 1983.

L. I. Zaichik and V. M. Alipchenkov. Collision rates of bidisperse inertial particles in isotropic turbulence. *Physics of Fluids*, 18:035110–1–035110–13, 2006.

L. I. Zaichik, V. M. Alipchenkov, and A. R. Avetissian. Modelling turbulent collision rates of inertial particles. *International Journal of Heat and Fluid Flow*, 27:937–944, 2006.

T. Zwinger. *Dynamik einer Trocken-Schneelawine auf beliebig geformten Berghängen*. PhD thesis, Vienna, Austria, 2000.

www.ingramcontent.com/pod-product-compliance
Lightning Source LLC
Chambersburg PA
CBHW021059210326
41598CB00016B/1257